Collaborative and Humanoid Robots

Edited by Jesús Hamilton Ortiz and Ramana Kumar Vinjamuri

Published in London, United Kingdom

IntechOpen

Supporting open minds since 2005

Collaborative and Humanoid Robots
http://dx.doi.org/10.5772/intechopen.91603
Edited by Jesús Hamilton Ortiz and Ramana Kumar Vinjamuri

Contributors
Bijan Paul, Amrita Ganguly, Hyun-Joon Chung, Arbnor Pajaziti, Xhevahir Bajrami, Gazmed Pula, Ashwin Dani, Iman Salehi, Ghananeel Rotithor, Ivan Nikolov Chavdarov, Sally Whelan, Dympna Casey, Dang Xuan Ba, Joonbum Bae, Chandan Choubey, Jyoti Ohri, Bülent Özkan, Jesus Hamilton Ortiz, Diego Saavedra Lozano, María Fernanda Díaz Velasquez, Javier F. Castillo

First published in London, United Kingdom, 2021 by IntechOpen
IntechOpen is the global imprint of INTECHOPEN LIMITED, registered in England and Wales, registration number: 11086078, 5 Princes Gate Court, London, SW7 2QJ, United Kingdom
Printed in Croatia

British Library Cataloguing-in-Publication Data
A catalogue record for this book is available from the British Library

Additional hard and PDF copies can be obtained from orders@intechopen.com

Collaborative and Humanoid Robots
Edited by Jesús Hamilton Ortiz and Ramana Kumar Vinjamuri
p. cm.
Print ISBN 978-1-83968-739-6
Online ISBN 978-1-83968-740-2
eBook (PDF) ISBN 978-1-83968-741-9

For EU product safety concerns:
IN TECH d.o.o., Prolaz Marije Krucifikse Kozulić 3, 51000 Rijeka, Croatia, info@intechopen.com or visit our website at intechopen.com.

Meet the editors

Jesús Hamilton Ortiz has a Ph.D. in Computer and Telecommunication Engineering. He is an international reviewer, editor, scientist, and entrepreneur in computer, telecommunication, and aerospace engineering. Dr. Ortiz has edited approximately nine books on ad hoc networks, telecommunication networks, wearables, Industry 4.0, mobile networks, mobile computing, and more. He is an associate editor of the journal IEEE Access. He is currently working on drone swarms, flying ad hoc networks (FANETs), bio-inspired algorithms, urban air mobility, air taxis, artificial intelligence, machine learning, the Volocopter, big data, blockchain, business intelligence, vertical–horizontal integration, and 5.0 society, and different disruptive technologies including cobots and humanoid robots.

Ramana Vinjamuri received a Ph.D. in Electrical Engineering with a specialization in dimensionality reduction in control and coordination of human hand from the University of Pittsburgh, Pennsylvania, in 2008. From 2008 to 2012, he worked as a postdoctoral research associate in the field of Brain-Machine Interfaces (BMI) to control prostheses at the School of Medicine, the University of Pittsburgh, where he received the Mary E Switzer Merit Fellowship from the National Institute on Disability, Independent Living, and Rehabilitation Research (NIDILRR) in 2010. From 2012 to 2013, he worked as a research assistant professor in the Department of Biomedical Engineering, Johns Hopkins University, Baltimore, Maryland, in the area of neuroprosthetics. He also worked as an assistant professor in the Department of Biomedical Engineering, Stevens Institute of Technology, Hoboken, New Jersey, from 2013 to 2020. He holds a secondary appointment as an adjunct assistant Professor at the Indian Institute of Technology, Hyderabad, India. He is currently an assistant professor in the Department of Computer Science and Electrical Engineering, University of Maryland, USA. In 2018, Dr. Vinjamuri received the Harvey N Davis Distinguished Teaching Award for excellence in undergraduate and graduate teaching. He also received the National Science Foundation (NSF) CAREER Award in 2019 and an NSF Industry-University Cooperative Research Centers (IUCRC) Planning Grant in 2020. His other notable research awards are from the US-India Science and Technology Endowment Fund (USISTEF) and the New Jersey Health Foundation.

Contents

Preface

Humanoid and collaborative robots are being developed to achieve broader and higher-level cooperation with humans. It is expected that more effective functional coexistence between humans and high-level robots will be achieved, especially in Industry 4.0 and its respective pillars.

This interaction between cobots, humanoid robots, and humans will become increasingly natural for the end customer, both at the industrial and domestic levels. Both humanoid and collaborative robots will increasingly develop cooperative capabilities in specifically designed environments.

With artificial intelligence (AI) being one of the main pillars of Industry 4.0, humanoid robots and cobots are expected to become increasingly capable of predicting the emotions, facial expressions, and gestures that humans typically make when performing collaborative work.

AI and machine learning will hopefully learn and process larger amounts of data. This increased processing of large volumes of data forces humans to rethink what is and what is not unique to humans. All the conjecture and speculations of the past regarding technological advances show us that we have to rethink all our preconceived ideas about humanoid robots, cobots, and AI, as this disruptive technology is changing all our technological concepts.

With Industry 4.0 there are new challenges in the production chain, as workers must actively collaborate with humanoid robots and cobots in an environment connected by 5G networks. Similarly, workers will have to be prepared to use body equipment and wearable devices to improve their safety. In addition, technological training will be a priority to work in this industry, since AI will be a fundamental part of both the production chain and management.

Some of the pillars of the intelligent industry include safety, AI, and the personalization of products and services. Humanoid robots, collaborative robots, and robotic wearables will be increasingly intelligent with a great increase in machine learning both in the production chain and at the domestic level.

Given the importance of AI, security, and personalization, perhaps the entire economic model will have to change to a more sustainable model. Therefore, it will be a priority to integrate the production chain of a smart industry with a circular economy model to minimize the negative impact on our planet.

Regarding human-robot interaction, humanoid robots, and cooperative robotics, we can observe a disruptive change at the technological level, especially at the level of industrial safety. The use of collaborative robots, humanoid robots, and robotic wearables allows more efficient production systems in human-robot interaction, thinking about the development of different daily tasks both in an industrial plant and at a domestic level.

The role of the human being in a smart factory is changing rapidly. Human workers must adapt to these new systems by acquiring and improving their skillsets. Therefore, it is important to equip them with new skills in disruptive technologies that will give them the necessary tools to work with collaborative robots, humanoid robots, and robotic wearables. For this, it is necessary to update the entire training model and the academic programs of different institutions focused on the teaching-learning process.

The concept of collaboration in the industry is fundamental as it integrates a group of people and robots (specialists, experts, or connoisseurs) performing an activity jointly, decentralized, and simultaneously to achieve a common goal. This concept must be constantly evaluated objectively, always taking into account the technological changes that are being generated at the level of innovation and research. Some of the advantages of the collaborative concept include:

- Multidisciplinary teams: formed by members with different skills and knowledge that contribute to improving production processes

- Increased productivity: the goal is to achieve greater efficiency in more agile environments

- Information storage and centralization: the goal is to achieve greater efficiency in more agile environments

Some detractors of disruptive technologies may think that robotics will put people out of work, increasing unemployment and causing a greater economic and social crisis. Perhaps the new technological models demand a change in the methods of production, education, training, and so on. It is important to mention that the relationships between humans and machines as well as between humans themselves will inevitably change. In addition, it will be necessary to migrate to a circular economy model, different from the linear economy model, which is not sustainable with the environment.

This book is the result of the work of people dedicated to research, academics, and applied industry. It is a useful resource for those readers interested in new contributions in robotics and related fields.

Jesús Hamilton Ortiz, Ph.D.
Closemobile R&D, and UNAD,
Madrid, Spain

Ramana Kumar Vinjamuri
University of Maryland,
Baltimore County, USA

Section 1

Collaborative Robots

Chapter 1

COBOTS in Industry 4.0: Safe and Efficient Interaction

Javier F. Castillo, Jesús Hamilton Ortiz, María Fernanda Díaz Velásquez and Diego Fernando Saavedra

Abstract

Cyberphysical systems will have a great development with the digital transformation known as industry 4.0. Cyberphysical Systems systems are devices that integrate capabilities to control and interact with a physical process. Among these are the Cobots, robots that perform tasks directly in conjunction with humans within a shared or nearby space. Safety is a fundamental issue when talking about Cobots, because there are requirements in terms of materials and design, kinetic limitations, and the implementation of sensors and algorithms that guarantee a safe workspace. Therefore, the potential risks of Cobot applications within the boom of industry 4.0 in cyberphysical systems are presented. Defining the fields of: Safety inspections; Routes and algorithms to avoid obstacles; Human-machine interfaces; Humans and Industry 4.0; Cyber security. Framed within the human-machine standards and protocols, within the safety functions and performance of a Cobot (ISO 10218-1 and ISO 10218-2 and ISO TS 15066). In this chapter, we discuss the different problems that are in the application of Cobots, in conjunction with different proposals for improvement and aspects to consider.

Keywords: Cobots, Industry 4.0, Safety, security, cybersecurity

1. Introduction

Contextualizing, industry 4.0 is the integration of cyber-physical systems in manufacturing and logistics processes, the use of the internet of things and services in industrial processes [1]. Creating significant changes in the manufacturing industry, in consumer behavior and in the way of doing business, through technologies that work together, such as additive manufacturing, 3D printing, reverse engineering, big data and analytics, artificial intelligence, autonomous robots that when working together, cause [2].

The term industry 4.0 was mentioned in 2011 at the Hannover fair. The "4.0" refers to a fourth industrial revolution due to the cyber-physical production systems that integrate the real and virtual [1]. Initially, the first Industrial Revolution, between the XVlll and XlX centuries was based on introducing mechanical production equipment powered by water vapor. The second Revolution in the twentieth century allowed mass production, thanks to the division of tasks and the use of electrical energy, which facilitated the manufacture of products for mass consumption. At the end of the twentieth century there was the third revolution based on

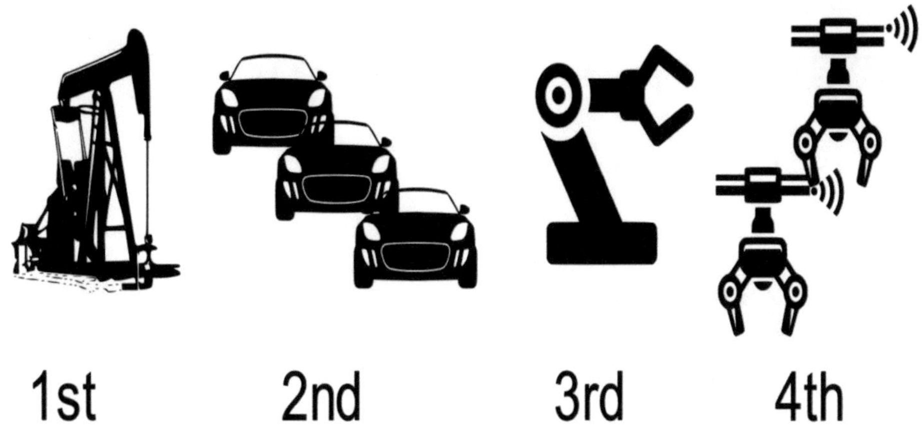

Figure 1.
Four industrial revolutions.

electronics and computer science to make automated production possible. Finally, industry 4.0 emerges as a new industrial revolution incorporating the new technological advances of the time [3].

In **Figure 1**, it is observed how the 4th industrial revolution, has a presence with robotics, intelligent technology and interconnectivity, thus giving way to collaborative robotics.

2. Cyber-physical systems

Cyber-physical Systems (SCF) incorporate and coordinate physical processes with computational elements, communication networks and remote storage of information, both interact at different scales of space and time, with different operation processes and configuration and reconfiguration resources. In addition, SCF works through "systems of systems" that interact and communicate with each other, through networks and software.

The key functions of Cyber-Physical Systems in the industry are the development of sensors and actuators to perform control actions, monitoring, generation of new knowledge, self-learning and reconfiguration based on the condition of the process. The mobile internet and the internet of things (IoT) have allowed to extend the adaptability, capacity, resilience, scalability, usability, and security of current industrialsystems. For example, wind-solar energy systems are an application of Cyber-Physical systems in industry, these wind and solar farms have their physical share while the data is obtained through monitoring sensors. The data is transferred through communication networks and processed by software in operation centers to monitor and control the physical environment, and thus obtain the most benefit from renewable sources. Also, CFS have been applied in different fields such as in the manufacture of intelligent robots, to perform human welfare services, in medical or healthcare equipment, vehicular and transport systems, surveillance systems, smart cities and video games.

The SCF contributed significantly to the industrial revolution, because in the former there were no intelligent factories where components, people and production systems communicated with each other through a network and their production was autonomous. In today's industry, CFS allow the monitoring of the condition of machines or processes in real time, the detection of anomalies, the predictions of failures, the cloud as a service increasing the realization of more productive, sustainable and efficient factories [4].

3. COBOTs

With the rise of the industry due to the introduction of cyber-physical systems (SCF) and the Internet of Things (IoT) within manufacturing and automation systems. SCF represent self-controlled physical processes, with strict network capabilities and efficient interfaces for human interaction. The interactive dimension of SCF reaches its maximum when defined in terms of natural human-machine interfaces (NHMI), i.e. those that reduce the technological barriers required for interaction [5].

A collaborative robot or Cobot is defined according to ISO/TC 299 as a robot designed to interact directly with humans in a defined collaborative space. Cobots are systems or robots of an industrial nature, designed to work in robot-human environments, that is to say that it is in contact with human operators, which allows the operator's work to be complemented with the characteristics of the machine. The automotive industry, surgery, training devices and rehabilitation are some of the applications in which Cobots have been used.

The industry is currently looking to take advantage of mass customization, a task that industrial robots pre-programmed for large-scale production are not focused on, so Cobots aim to address the changes that occur in large-scale customization processes. Companies are using them because they can work alongside humans on the same assembly lines using small spaces, the necessary security measures are much less than in simple robots, they are very flexible to work on short consignments without having to invest much time in programming them and they are ideal to replace operators in repetitive or poorly ergonomic tasks.

The digitization and intelligence of the manufacturing process is the need of today's industry. Rapid advances in manufacturing technologies and applications in industries help increase productivity. Industry 4.0 as a new level of organization and control over the entire value chain of the product life cycle; it is geared towards the requirements of increasingly individualized customers. Industry 4.0 is still a hypothetical concept, but with real development that includes the Internet of Things, industrial Internet, smart manufacturing, and cloud-based manufacturing. Industry 4.0 refers to the strict integration of the human being in the manufacturing process to have a continuous improvement and focus on value-added activities and avoid waste [6].

The collaborative robotics of industrial automation has adapted to the context of rapid industrial development. The collaboration of humans and robots in close proximity in a single workspace is a novel concept of Industry 4.0. Research topics in the field of industrial robotics include safety in human-machine interaction systems, due to the close collaboration between humans and robots many of the problems of industrial robotics are associated not only with technical issues, but also with social aspects. They also include the optimization and automation of production through the introduction of robotic solutions, with which it is expected that in the future Cobots will become the basis of a large commercial development [7].

Robots are tough, fast and very accurate machines that can complete their tasks faster, with better quality and for a lower price than humans. However, some operations must be adapted to real events, but robots are not able to think, execute commands and performpre-learnedmovements, being limited by their programming. Manipulation robots are usually designed to have six or seven degrees of freedom (axes of motion), while the upper extremity of the human body has around thirty. This results in another precise manipulation limitation with a wide range of motion.

With mass custom production becoming mainstream, agile manufacturing strategies have been adopted by small and medium-sized industries (SMEs), driving the use of collaborative robots in today's factories. The main challenges in the adoption of Cobots in the industry are the lack of a highly skilled workforce to program the robot to perform complex tasks and the integration of robotic systems to other smart devices in the factory. In addition, teaching and simulation by non-robotics experts of many collaborative robot systems is a great challenge, because these systems are designed to be programmed by experts and not by ordinary workers [8].

The main goal ofhuman-robotcollaboration (HRC) is to create an environment for safe collaboration between humans and robots. There is an area between manual manufacturing and fully automated production where a human worker comes into contact with the machine. This area has many limitations due to security restrictions. The machine can be in automatic operation only if the operational staff is outside their workspace. Collaborative robotics establishes new opportunities in cooperation between humans and machines, where staff share the workspace with the robot helping with non-ergonomic, repetitive, uncomfortable, or even dangerous operations. The robot monitors its movements by using advanced sensors so as not to limit, but mainly not to endanger the worker [9].

3.1 Programming a Cobot

The programming process of a Cobot involves the ability to understand the state of the environment and perform actions that advance the system towards a planned goal of collaboration. The programming characteristics of Cobots identified are [10]:

- Communication: an operator controls an Cobot through a communication channel that can be verbal (speech) or nonverbal. The programmer's offline function is to program and define possible actions of Cobot and the underlying motion control.

- Optimization: Important aspects of an Cobot senvironment, such as obstacles and tool positions, are mathematically modeled based on the actions of the Cobot. Those form cost functions that are optimized to generate desirable performance. The Cobot program can be carried out to minimize the operator's workload, energy consumed and lost time, or maximize physical comfort and confidence, product quality, etc.

- Learning: An Cobot Clearns a skill like what a human would, for example, by observing demonstrations, trial and error, receiving feedback, and asking questions. The role of a programmer is to design the learning algorithm and provide initial data for the Cobot to learn. That could be in the form of demos, trial-and-error iterations, training data, etc.

3.2 Cyber security

Digitization strategies in cyberphysical production systems (CPPS) are one of the key factors in Industry 4.0. Its integration into a hyperconnected system facilitates the production of goods and services. In addition, these industries are characterized by automation, as well as unmatched levels of data exchange across the value chain. The topic not only addresses data preparation, real-time data processing, big data analysis, visualization, and machine interface design, but also cybersecurity. In particular, unauthorized access to protected data (personal or business) or unauthorized control of production facilities involve risks in terms of digitisation, with digitisation having an impact on security. Cybersecurity risks are crucial as the prevalence of these information and operation technologies has changed the appearance of cyber threats. Addressing the premises and realities of cybersecurity in Industries 4.0 and 5.0 is crucial. The risk mitigation strategies provided by various organizations are crucial to reducing risks. Given the gaps and vulnerabilities generated by interconnections, cybersecurity is vital for the advancement of digital industrial transformation [11, 12].

3.3 Standards

Industrial regulations that incorporate the risks related to the use of collaborative robots by workers include the international standard ISO 10218 and the Technical Specification ISO/TS 15066: 2016, the American ANSI/RIA R15.06, the European EN 775 that is adapted from ISO 10218, and standards such as the Spanish UNE-EN 755 adapted from EN 755 by the Spanish Association of Standardization and Certification. To prevent accidents, the selection of a safety system should be based on the analysis of the aforementioned risks. Commonly in the past, security systems have separated workspaces from robots and humans. An instance of this separation was reflected in the UNE-EN 755: 1996 standard. It stated that sensor systems should be incorporated to prevent people from entering a hazardous area, where the operational state of the robotic system could have caused dangers to workers. According to traditional standards, authorized personnel can only be inside the robot workspace if the robot is not in automatic mode.

The latest update of ISO 10218-1 and ISO 10218-2 provide details on collaborative work requirements and typologies of cooperation tasks. The first includes, for example, start-up controls, operation of the safety control system, motion braking, speed control, while the second includes, for example, manual guidance, interface window and cooperative workspace. The international standard ISO: 8373–2012, specifies the vocabulary used in relation to robots and robotic devices. New terms involved in the development of new collaborative tasks in industrial and non-industrial environments, such as human-robot interaction and the service robot, are defined, as well as more established terms, such as robot and control system. The recent Technical Specification ISO/TS 15066: 2016, attempts to further specify human-robot collaboration by complementing the requirements and guidance set out in ISO 10218.

3.4 Rules and types of cooperation

Iso EN 10218 for robots and robotic devices defines four basic types of HRC. For some types of cooperation, the use of special collaborative robots with integrated sensors is required. Other types of applications have a conventional robot with improved sensors and controls.

The supervised *stop* with safety rating is the simplest type of collaboration. There are applications in which the robot share's part or all its workspace with an operational team. If a worker appears in the robot's work area, the machine stops and remains on hold until the man leaves. In the shared area, the robot and the operator can get to work, but not at the same time.

During the operator's manual control process, the robot's load is compensated to maintain its position. The operator can move freely with the manipulator in space without exerting a force majeure. The human being comes into direct contact with the machine, but the movement is not initiated by the robot, only controlled by the operator. For safety reasons, the speed of the robot is reduced and updated with safety functions. The robot must be equipped with a measuring device to monitor the impact load. Some robots have sensitive elements (torque sensors) embedded directly in the joints. For this type of collaboration, it is also possible to use a standard robot. The robot must be equipped with a sensor that can detect external loads. This sensor is placed on the robot's wrist between the output interface and the final effecter. Measure and evaluate the load and verify the robot's compliance.

With speed control and separation, the robot cell work area is divided into different areas. These areas are inspected with scanners or a vision system. In areas beyond the range of the manipulator, where the operator is not in contact with the robot, but could run the risk of a manipulated object falling, the robot slows down to a safe speed. If the robot's workspace is interrupted, the robot will stop. If these two areas are free, the robot can work with the maximum parameters. The speed and position of the robot are continuously monitored. A recommended application may be a workplace where the robot works with the maximum parameters, but the operator must enter the area at a certain time. For example, for logistical reasons, to place or remove the product.

Force and power limitation is a type of cooperation where special collaborative robots are needed. The movement parameters of the robots are controlled with high precision and even a small deviation from the actual position compared to the programmed one can be determined. Accurate encoders with high resolution allow the robot to control its speed and position with high precision. The forces and pairs are measured and evaluated with sensitive torque sensors in the robot joints, checking the electric current consumed in the actuators, measuring the reactions transmitted to the ground or implementing tactile sensors. Therefore, the robot is able to identify the impact on an obstacle, analyze it in a short time and react. The robot can brake after the collision and stop immediately, alternately moving in the opposite direction in the opposite direction to decrease the energy of the impact as much as possible.

3.5 Efficiency in interaction

The field of collaborative robot research is constantly evolving. There is no particular approach to increasing the efficiency of intelligent human-robot robotic systems. This is due to the complexity of ensuring a coordinated interaction between the two parties [12]. For this reason, adaptive control systems for the robot, intelligent human-robot interface, sensor systems and information processing algorithms, the basic elements of robotic and mechatronic systems, efficient forms of interaction of the robot with the external environment, bionic and biomedical technologies in robotic solutions, control of multi-agent robotic systems with secure and reliable communication, continues to be constantly developed.

Figure 2 shows the difference between interactions and presence with robots. In collaborative robotics, the task or work objective is performed in the company of an operator or worker of the company. In an industrial process where tasks are automated, the machine is the one who performs the tasks individually or sharing objectives with other machines. Finally, manufacturing is done directly by people; this is because they are craftsmanship or they are only achievable manually, although somewhere in the process, robots can get involved and become collaborative robotics.

With the aim of identifying the factors of effectiveness of the implementation of robotic solutions according to the use cases in companies that have introduced robotics in their production processes. Based on the opinions of experts with experience and knowledge of the robotics market, the following factors were formed in descending order of importance in the article [12]:

1. Increased productivity.

2. Quality improvement.

3. Reduction of labor costs.

4. Elimination of dangerous operations.

5. Increase production flexibility.

However, in the modern trend of implementing collaborative robotic solutions in a shared workspace and in everyday human activities focuses on the person and the effectiveness of their direct contact [13].

Collaborative robots successfully achieve their goal of reducing ergonomic risks and improving employee safety. In addition, important levels of productivity improvement were identified, and the collaborative robot managed to stabilize the behavior of the assembly station, since its performance and pace of work are more consistent than those of a human operator.

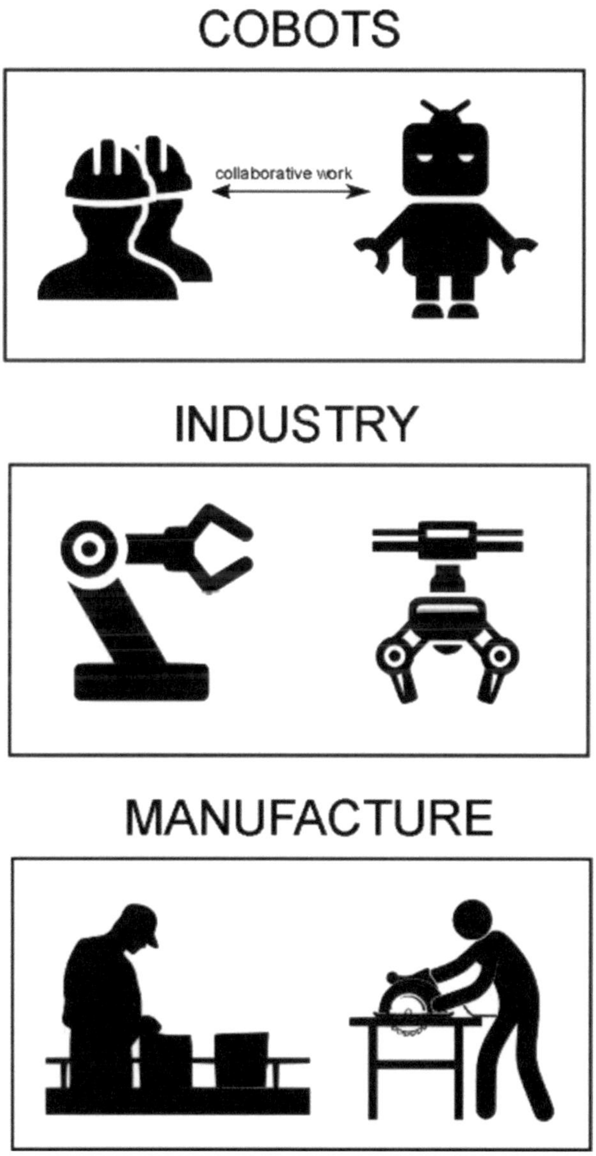

Figure 2.
Comparison of cobots, automated industry and manufacturing.

1. Perform a risk analysis associated with the use of the collaborative robot: Two important sources of risk were identified: mechanical entrapment and shocks, which are caused by the movements of the collaborative robot. In this sense, it is important to implement the necessary security measures to minimize the incidence and severity of these risks.

2. Validate the integration of the collaborative robot with the assembly station: The use of the collaborative robot on the production line was validated according to the evaluation and validation procedures of the company.

3. Train technicians and operators in the maintenance and collaborative operation of robots, respectively: The two trainings were given and documented in accordance with the company's policies [14].

3.6 Risks

In the current industry 4.0 the tools are divided into two groups, one which bases its operation on capturing and processing data, and the physical and tangible tools [15].

The latter can present a series of risks such as:

- Risk in the planning and organization by the company, for the acquisition of a new technology

- Psychosocial risk, in terms of psychological, emotional, and social involvement of the team

- Safety, hygiene, and ergonomic risks generated by staff interaction with new technologies.

- Cybersecurity, in terms of vulnerability of companies' information due to data exchange and technological connectivity.

Safety: Mechanical contacts that generate physical damage because I) adjustments, programming, or tests of the robot, II) access to restricted areas during automatic operation, III) During cleaning or maintenance, IV) During collaborative work. As well as electrical risks, when contacting active parts, by poorly insulated elements, incorrect voltage during maintenance. Thermal risk burns and danger from flammable atmosphere. Projection of materials, sparks, or particles.

Safety: Mechanical contacts that cause physical damage through I) configuration, programming or testing of the robot, II) access to restricted areas during automatic operation, III) during cleaning or maintenance, IV) during collaborative work. In addition to electrical hazards from contact with living parts, poorly insulated elements, incorrect voltage during maintenance. Thermal hazard burns and danger of flammable atmosphere. Projection of materials, sparks, or particles.

Hygiene: Exposure to I) Vibration that endangers the operation of the cobot or the health of the operator, II) Ionizing and non-ionizing radiation, III) Materials or chemical elements harmful to health, IV) Noise, V) Magnetic fields.

Ergonomics: Due to repetitive movements that must be done with the arm or hand, prolonged forced postures, overexerted by the weight that is loaded, postural changes by the reduced space.

Psychosocial: Problems in adaptability with COBOT, change in work rhythm, overload, and mental fatigue.

Workspace	Action	Algorithms	Hardware and sensors
separate	Restriction on staff	none	signage and delimitations. Light, acoustic, optical, etc. indicators.
separate	Change in robot behavior. Stop the robot or slow it down.	control algorithms	Combination of passive systems and security assets. Interlocking device, proximity, and touch sensors.
shared	Quantification of the level of risk per collision	none	Estimation of pain tolerance, assessment of the level of risk. Human arm emulation system, Standard car crash test
shared	Minimize risk from collision or deliberate contact	none	technical mechanical compliances, lightweight structures. Viscoelastic coating, elastic absorption systems, use of carbon and aluminum fibers.
shared	Minimize risk from collision or deliberate contact	Safety strategies for collision detection	Touch sensors, proprioceptives, encoders, force sensors, optical and RGB-D sensors.
shared	obstacle avoidance, stopping or changing trajectory, speed, force	preclusion analysis strategies	Motion capture systems, local information, computer vision, distance sensors, RGB-D. Capacitive sensors, ultrasound, laser, IR, cameras, etc.

Table 1.
Human-robot symbiosis [16].

It is important to keep in mind that, in collaborative robotics, there are different ways to interact or work with robots, in the **Table 1** is described these interactions. It may be the case in which the machine is isolated from the operator, this for the safety of the latter. Another case is when both the machine and the operator are isolated; process in which greater security is required for the person. And finally, when the machine needs direct handling or is not risky for the operator, then there is no separation between these. **Figure 3** depicts these separations.

3.7 Applications

The prevention of collisions between humans and robots is fundamental in collaborative robotics and in the framework of Industry 4.0. It plays an important role in meeting safety criteria, as people and machines work side by side in an unstructured and time-varying environment. Autonomous guided vehicles (VGA) implement techniques for navigating through mapping, location, route idealization, and route tracking. This technique allows that, if there are any obstacles in the way, the transport does not have to stop. Instead of stopping the machine, by using point-to-point motion logic, a tool was implemented to avoid obstacles. Often, the only hardware available in VGA is PLCs. This hardware limitation is a prominent feature, as not all computers support large computing capabilities [15].

Some applications of Cobots expressed in [17] are:

- Industrial application of *Universal Robots,* Collaborative robots on BMW assembly lines. Robots are used on the production line to wrap a layer of protective film over electronic components inside a door, which can cause repetitive strain injuries to workers when executed manually.

- Audi's human-robot cooperation in production processes is based on the "*PART4you*" robot. It incorporates a camera and suction cup to help human workers pick up the components of the boxes and pass them on to the assembly workers, without safety barriers, at the right time and in an ergonomically ideal position.

- *KUKA's* collaborative robots are used on an assembly line to help workers install shock absorbers: Instead of using a heavy damper installation tool, workers have the robot automatically lift and place the damper on the wheel arch before pressing a button to install the component.

- Universal Robots' robotic arms are also used at the Volkswagen factory, where they handle delicate incandescent spark plugs on cylinder heads, allowing for ergonomic design of the factory's workplace, where the employee can complete the task of repairing spark plugs.

- *ŠKODA* production employees work together with robots on high-precision tasks such as inserting the piston of the gear actuator, which is one of the most delicate processes in the manufacture of transmissions.

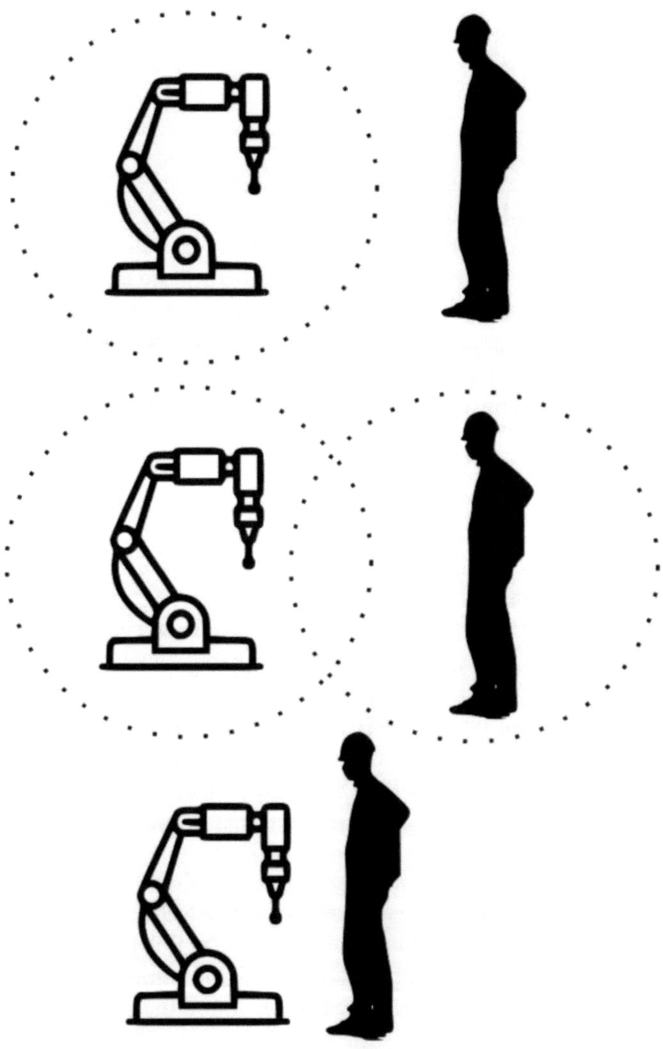

Figure 3.
Types of interaction with the robot.

3.8 Future applications

Although existing industrial robots can work in challenging environments, perform high-precision tasks, and help improve and increase productivity, most of this is still driven by pre-designed robot commands and programs. Given labor costs and intense competition, there is a huge shortage of autonomous and intelligent robots and physical cyber systems capable of perceiving and deciding in the next application of Industry 4.0. These intelligent robots can analyze their tasks by selecting the right tools, planning their movements, and performing the right operations in the same way that a trained human worker would [17–21].

With the constant advances that occur in robotics and with the demand for new needs that occur in industries, the presence of collaborative robots is increasingly noticeable, especially in those processes that are repetitive and can be divided between a machine and a person.

Author details

Javier F. Castillo[1], Jesús Hamilton Ortiz[2*], María Fernanda Díaz Velásquez[1] and Diego Fernando Saavedra[1]

1 Facultad de Ingeniería, Universidad Santiago de Cali, Colombia

2 MOABIT, Madrid, Spain

*Address all correspondence to: jesushamilton.ortiz@gmail.com

References

[1] C. J. Bartodziej, "The concept industry 4.0," in *The concept industry 4.0*, Springer, 2017, pp. 27-50.

[2] C. B. Y. Cortés, J. M. I. Landeta, and J. G. B. Chacón, "El entorno de la industria 4.0: implicaciones y perspectivas futuras," *Concienc. tecnológica*, no. 54, pp. 33-45, 2017.

[3] J. L. del Val Román, "Industria 4.0: la transformación digital de la industria," in *Valencia: Conferencia de Directores y Decanos de Ingenier{\'\i}a Informática, Informes CODDII*, 2016.

[4] A. Villalonga Jaén, F. Castaño Romero, R. Haber, G. Beruvides, and J. Arenas, "El control de sistemas ciberf{\'\i}sicos industriales: revisión y primera aproximación," in *XXXIX Jornadas de Automática*, 2018, pp. 916-923.

[5] M. A. R. Garcia, R. Rojas, L. Gualtieri, E. Rauch, and D. Matt, "A human-in-the-loop cyber-physical system for collaborative assembly in smart manufacturing," *Procedia CIRP*, vol. 81, pp. 600-605, 2019.

[6] S. Vaidya, P. Ambad, and S. Bhosle, "Industry 4.0--a glimpse," *Procedia Manuf.*, vol. 20, pp. 233-238, 2018.

[7] R. Galin and R. Meshcheryakov, "Automation and robotics in the context of Industry 4.0: the shift to collaborative robots," in *IOP Conference Series: Materials Science and Engineering*, 2019, vol. 537, no. 3, p. 32073.

[8] V. V. Nair, D. Kuhn, and V. Hummel, "Development of an easy teaching and simulation solution for an autonomous mobile robot system," *Procedia Manuf.*, vol. 31, pp. 270-276, 2019.

[9] A. Vysocky and P. Novak, "Human-Robot collaboration in industry," *MM Sci. J.*, vol. 9, no. 2, pp. 903-906, 2016.

[10] S. El Zaatari, M. Marei, W. Li, and Z. Usman, "Cobot programming for collaborative industrial tasks: An overview," *Rob. Auton. Syst.*, vol. 116, pp. 162-180, 2019.

[11] T. Komenda, G. Reisinger, and W. Sihn, "A Practical Approach of Teaching Digitalization and Safety Strategies in Cyber-Physical Production Systems," *Procedia Manuf.*, vol. 31, pp. 296-301, 2019.

[12] A. Clim, "Cyber security beyond the Industry 4.0 era. A short review on a few technological promises," *Inform. Econ.*, vol. 23, no. 2, pp. 34-44, 2019.

[13] Iberdrola, "'Cobots': Los nuevos robots colaborativos." [Online]. Available: https://www.iberdrola.com/innovacion/cobots-robots-colaborativos. [Accessed: 09-Jun-2021].

[14] A. Realyvásquez-Vargas, K. C. Arredondo-Soto, J. L. Garc\'\ia-Alcaraz, B. Y. Márquez-Lobato, and J. Cruz-Garc\'\ia, "Introduction and configuration of a collaborative robot in an assembly task as a means to decrease occupational risks and increase efficiency in a manufacturing company," *Robot. Comput. Integr. Manuf.*, vol. 57, pp. 315-328, 2019.

[15] S. Salimbeni and D. Mamani, "Marco de referencia para la incorporación de Cobots en l{\'\i}neas de manufactura," *Podium*, no. 38, pp. 159-180, 2020.

[16] S. Robla-Gómez, V. M. Becerra, J. R. Llata, E. Gonzalez-Sarabia, C. Torre-Ferrero, and J. Perez-Oria, "Working together: A review on safe human-robot collaboration in industrial environments," *IEEE Access*, vol. 5, pp. 26754-26773, 2017.

[17] V. Villani, *F. Pini*, F. Leali, and C. Secchi, "Survey on human--robot

collaboration in industrial settings: Safety, intuitive interfaces and applications," *Mechatronics*, vol. 55, pp. 248-266, 2018.

[18] P. S. d. l. Reyes, Gestión de la producción en la Industria 4.0 (tesis de grado), Sevilla: Dpto. de Organización Industrial y Gestión de Empresas I - Escuela Técnica Superior de Ingeniería, 2020.

[19] F. Cassioli, G. Fronda y G. Fronda, «Human–Co-Bot Interaction and Neuroergonomics: Co-Botic vs. Robotic Systems,» Frontiers in Robotics and AI, 2021.

[20] International Federation of Robotics - IFR, «IFR presents World Robotics Industrial Robots,» 2019. [En línea]. Available: https://ifr.org/ifr-press-releases/news/robot-investment-reaches-record-16.5-billion-usd.

[21] S. Bragança, E. Costa, I. Castellucci y P. M. Arezes, «A Brief Overview of the Use of Collaborative Robots in Industry 4.0: Human Role and Safety,» Occupational and Environmental Safety and Health, vol. 202, pp. 641-650, 2019.

Chapter 2

Examining Social Robot Acceptability for Older Adults and People with Dementia

Sally Whelan and Dympna Casey

Abstract

Social robots that aim to support the independence and wellbeing of older adults and people with dementia are being introduced into dementia care settings. However, the acceptability of robots varies greatly between people and the rate that robots are deployed into practice is currently low. This chapter defines robot acceptability and provides an overview of theoretical technology acceptance models. It reviews the empirical literature and identifies the individual and contextual factors that impact acceptability in relation to the needs of older adults and people with dementia, focusing on what potential robot users need to motivate them to accept robots into their everyday lives. Then the literature is discussed in the light of current discourses in gerontology, recommending what is needed to increase the acceptability of robots. The capacity of robots, to communicate in a human-like way needs to increase and robots need to be designed with in-depth end-user collaboration, to be person-centred and deployed in ways that enhance the strengths of people with dementia. Guidance for good practice in participatory design is provided. Longitudinal research that uses triangulated data from multiple sources. is recommended to identify the needs of individuals, significant others, and wider contextual factors.

Keywords: Social Robots, Acceptability, Dementia, Alzheimer's Disease, Assistive Technology

1. Introduction

Dementia is a progressive neurological syndrome that causes cognitive degeneration, memory loss, alterations in personality and mood, difficulty performing everyday tasks and communication challenges [1]. As more people live into old age it is anticipated that worldwide, dementia will affect 66 million people by 2030 and 115 million by 2050 [2].

Social robots are artificial intelligence systems that are equipped with social and communicative abilities [3]. They differ from other robots as they are designed to interact using the social rules of human behaviour [4]. Technologies have been identified as key enablers for policies that aim to support aging in place and the sustainability of welfare states [5]. Currently, social robots are being designed to support the resilience [6] and independence of people with dementia and older adults (OA). They are also being developed to supplement the care that people with dementia receive from human caregivers, as diversion therapy and entertainment [7], to reduce loneliness [8], and to perform cognitive screening tests [9].

Some OA and people with dementia positively evaluate using robots [6, 8, 10]. But a recent review involving n = 90 research studies found that the deployment of current assistive robots into healthcare practice is low [11]. Technology acceptance is an important predictor of the usage and adoption of technology [12].

Acceptance has been defined as the robot being willingly incorporated into the older person's life [13]. Robot acceptance is impacted by multiple interacting factors, that concern the individual, significant others, and wider society [14]. But the factors that impact robot acceptability need to be understood in more depth because individual people and groups greatly differ regarding how acceptable they find robots [15, 16]. A recent literature review that included studies investigating robots (n = 4) found there was low acceptance of robotic technologies amongst OA and people with cognitive impairment and their caregivers [17]. Korblet [18] also found that the acceptance of a mobile telepresence robot in a nursing home setting was lower in people with dementia (n = 5) than in people who visited the nursing home centre weekly (n = 3) and a third who had physical disabilities (n = 3).

This chapter provides an overview of the state of the art regarding technology acceptance theory and the empirical research on robot acceptance that involves OA and people with dementia. It also examines what is likely to increase the acceptability of robots in this context. The chapter proceeds in the following sequence. Firstly, the definition of robot acceptability is discussed and phases in technology acceptance are introduced in relation to robots as a new technology genre. It is highlighted that social robots have different features in comparison to traditional technologies. These differences affect their acceptance and how acceptance needs to be conceptualised and investigated in practice. Secondly, examples of models of technology acceptance are presented and discussed in relation to their adequacy to explain and predict the acceptance of robots in the concept of OA and people with dementia. Then, empirical research is described that has examined the acceptance factors that are important for OA and people with dementia. The final section of the chapter focuses on what is likely to increase the acceptability of robots for OA and people with dementia and recommendations are made for the future design, development, and deployment of social robots. This final section introduces the concept of collaborative robots and it draws on research findings and literature that describes current discourses in gerontology and policy recommendations.

2. Examining the acceptability of robots for older adults and people with dementia

2.1 What is robot acceptability?

2.1.1 Definitions of technology acceptability

Definitions of technology acceptability vary and they establish acceptance in relation to other concepts like technology usability, adherence, and adoption. It can be defined as the intention to use technology [17]. Such a definition regards the actual usage of the technology as occurring after acceptance. Other definitions, for example, Heerink et al. [19], regard acceptance as including usage of the technology over a long period. Technology acceptance is regarded as a process [20] in which the user is involved in ongoing evaluation and re-evaluation of the technology and their decision whether to adopt the technology [4]. Ongoing decision-making is informed by information and experiences that approve discontinuance of use or confirm the initial decision to use the technology. Cruz [20] defines technology acceptance as a process and argues that acceptance only occurs when a technology

Phase	Key Features
Expectation	Before using, initial opinion for various information sources.
Confrontation	Re-evaluate ideas due to trying out or observing others for the first time.
Adoption	Initial serious user experiences
Adaptation	After 1 month, using according to personal needs and preferences.
Integration	After 2 months, meaningfully integrated and no longer noticed in everyday life.
Identification	After 6 months, is an expression of lifestyle or a group.

Table 1.
The Phases of Technology Acceptance (Adapted from de Graaf [4]).

sufficiently satisfies all the needs and requirements of its users. The concept of technology adoption has been distinguished from acceptance. De Graaf [4], who writes from the perspective of social robots for usage in the domestic home environment, defines adoption as being the initial decision to buy and start using the technology. The process of technology acceptance has six phases [4] that start with an individual becoming aware of a technology and end with the incorporation of that technology into everyday life to the extent that its functional purpose is exceeded and the individual becomes attached to it. These phases are summarised in **Table 1**) and described below.

2.1.2 Phases of technology acceptance

The expectation phase occurs when users form an initial opinion about a technology and an expectation about it. This opinion is formed by information they receive from the media, other people, and information they might seek out wanting to know more about the technology. When users try out or observe others encountering a technology for the first time, they enter the confrontation phase. Being exposed directly to the technology may cause the user to re-evaluate and adjust their expectations. The user enters the adoption phase when they use technology in a private environment and gain their first serious user experiences. This is followed by the adaptation phase when, after approximately a month, users have a broad idea of what the technology is about and they have encountered its flaws and features. As a result of exploring its features, users adapt the technology and use it according to their personal needs. Approximately two months after confrontation, during the integration phase, the technology becomes meaningfully integrated into the user's life, to the extent that it is not noticed by them and it has been personalised to their preferences. The final acceptance phase is the Identification phase. This occurs approximately six months after confrontation. Here the technology has more than a functional purpose. It can also then express a lifestyle and potentially be used to differentiate or connect groups of people. In order for robotic technologies to reach the integration and identification phases of acceptance, they need to offer users something that traditional technologies do not [4]. Before examining acceptance in relation to robotic technologies further, it is necessary to clarify how social robots differ from traditional technologies, and what this means for their acceptance.

2.1.3 Robots as a new technology genre

Robots are a relatively new genre of technology and they are not widely used in society. The level of robot technology acceptance changes over time and between contexts. Changes are impacted by the stage of the technology's development regarding its capacity and its level of diffusion in society [21]. The relatively little diffusion of robots in society currently impacts acceptability factors such as attitudinal beliefs about the technology. Lack of familiarity also impacts how at ease people are with a technology and their need for information in order to understand how it can be useful to them.

2.1.4 What is different about social robots?

The acceptance of robots is also impacted by the additional features that robots possess in comparison to traditional technologies. Robots are embodied devices that share the space with their users. Embodiment impacts how robots are perceived, and how people interact with robots and the type of relationship humans can build with them. OA prefer to interact with an embodied social robot rather than a computer screen [22].

Robots are also designed to behave and move more autonomously than traditional technologies. In addition, robots are designed with the intention to promote social interaction between themselves and the human user. These features impact their acceptability because the robot has a social presence and to be effective it must cause the human user to perceive it as a social identity. Variables including social presence and perceived sociability will be discussed further below in relation to the theoretical models that have used these concepts to explain and predict technology acceptance. Some examples of these models and how they have been used in research practice will now be presented.

2.2 Models used to explain and examine robot acceptability

2.2.1 Traditional technology models of acceptability

The Theory of Planned Behaviour was developed from the Theory of Reasoned Action [23]. This model proposes that people are influenced to behave in a certain way by making rational decisions about the personal and social outcomes that they anticipate as a result of their behaviour. This model may be helpful to explain the intention to use robots [4]. However, it cannot adequately explain the acceptance of robots, without being adapted. Because the decision to use robots can be impacted by emotional reactions rather than rational decisions [24].

The Technology Acceptance Model (TAM) [25] was derived to focus on the acceptance of computerised information systems in workplace contexts. TAM has been reported to explain 40% of variance in acceptance [26]. This model regards acceptance in terms of Intention to Use (ITU) the technology. Actual usage of the technology may or may not follow a potential user having ITU a robot. ITU is considered to be dependent on the user's attitudes towards the technology which is derived from their assessment of its Perceived Usefulness (PU) and on users' Perceived Ease of Use (PEOU). TAM was developed further [27] to include the impact of social influence, facilitating conditions, and habitual usage on technology acceptance.

The Unified Theory of Acceptance of Technology (UTAUT) offers a social psychological approach that can explain 70% of acceptance variance [27]. It uses constructs from eight previous theoretical models. It has four independent variables: performance expectancy, effort expectancy, social influence, and facilitating conditions. The independent variables affect the dependent variables: ITU and actual usage. Their effect on ITU and actual usage is moderated by gender, age, experience, and voluntariness of use. The UTAUT has been criticised as not being parsimonious and for combining highly correlated variables that provide an artificially high variance [4].

All the traditional acceptance models cannot adequately explain the acceptance of robots as social actors and the social aspects of human-robot interaction. They can also not fully enable the evaluation of pleasure-orientated factors that are necessary to motivate people to use robots in voluntary non-working environments [4, 28].

2.2.2 Acceptance models adapted for social robots

The Almere Model was developed using experiments (n = 4) that involved three robots, to test acceptance of social robots by OA (188) [19]. The Almere model extends and adapts the UTAUT. It has eleven constructs which are defined in **Table 2**. These constructs enable the measurement of acceptability to focus on aspects of the technology pertinent to social robots and the perception of humans towards them as autonomous embodied social entities. Heerink et al. [19] found that the model was strongly supported accounting for 59–79% of the variance in usage intentions and 49–59% of the variance in actual use. However, this was achieved using equation modelling on four separate databases without confirming their similarities [4]. It has also been suggested that users' beliefs about their self-efficacy to use and control the robot are underrepresented in the Almere model [18].

The Almere model has been used to explore levels of acceptance and engagement with a humanoid robot, Matilda [7]. This study involved people with dementia (n = 115) living in care homes (n = 4). It used a mixed-method longitudinal experience in which the reactions of participants were coded (n = 8304) according to emotional visual behavioural and verbal engagement measures. Participants used Matilda in three repeated 4–6 hour stages of field trials and as a result of the feedback received from participants, Matilda was further developed and improved.

The Model of Social Robot Acceptance was developed by expanding and adapting the theory of planned behaviour using some elements of the UTAUT and including factors relevant to robots [4]. This model utilises eight constructs that are summarised in **Table 3**. De Graaf's model regards users' attitudes as being comprised of both hedonic and utilitarian beliefs and it includes consideration of control beliefs and that people may use a robot, and any other assistive devise, due to habitual behaviour rather than making a rational decision to use it. This

Construct	Definition
Anxiety	Evoking anxious reactions when using the robot.
Attitude	Positive or negative feelings about the robot.
Intention to Use	The outspoken intention to use the robot over a long period.
Perceived Usefulness	The extent to which a user thinks a robot will be helpful.
Perceived ease of use	The degree to which the user believes they can use the robot.
Perceived enjoyment	Feelings of pleasure associated with the use of the robot.
Social Presence	The experience of sensing a social entity when interacting.
Perceived sociability	The perceived ability of the robot to perform sociably.
Trust	The belief the robot performs with personal integrity.
Perceived adaptability	The perceived ability of the robot to be adaptive.
Facilitating conditions	Factors in the environment that facilitate using the robot.
Social Influence	The user's perception of what other people think about them using the robot.

Table 2.
Constructs in the Almere Model (Adapted from Heerink et al. [29]).

Construct	Definition
Personal Norms	An individual's beliefs about using the robot, including privacy, trust, and attitude towards the robot.
Social Norms	A user's evaluation of the social consequences of using the robot. This includes social influence, media influence, and impacts on status.
Control Beliefs	User's beliefs about resources, opportunities obstacles affecting the use of the robot, including self-efficacy, previous experiences, prior expectations, personal innovativeness, safety anxiety, and the cost.
Utilitarian Attitudes	Includes perceptions of the robot's usefulness and how easy it is to use, including adaptability, embodiment, and robot personality.
Hedonic Attitudes	Evaluation of emotions, including enjoyment and pleasure that might arise from using the robot.
Habit	Behaviour resulting from the habitual use of the robot.
User Intention	The intention of the user to utilise and interact with the robot.
User Behaviour	The user interacting with the robot.

Table 3.
Constructs in the model of social robot acceptance [4].

model was tested by de Graaf [4] with respondents from an online questionnaire (n = 1248) of whom, 24.7% were aged over 60. This found that intention to use the robot increased when people believe: they have the skills to use the robot; that using it would be enjoyable; it would increase their status; when the robot would be less sociable; and cause less worry about privacy. However, only self-efficacy significantly impacted user intention to use the robot. De Graaf [4] concludes that the model is a useful guide to the identification of acceptance variables but that it needs further development and testing with other data sets. The Model of Social Robot Acceptance was used by Korblett [18] in the study examining telepresence robots that has been described above.

One limitation of applying the Almere model and the Model of Social Robot Acceptance to OA and people living with dementia is that they do not take account of disability and the fact that OA and people with dementia may have additional needs concerning technology usage. It has been argued that everyday and psychosocial functioning are better predictors of robot acceptance than chronological age per se [16]. The usage of technology is impacted by a person's level of cognitive ability which affects their psychomotor speed, domain knowledge, and visual memory [11]. People with dementia may also experience declining touch sensitivity and less ability to execute accurate and discrete movements [6, 30]. Learning new technologies can also be difficult with symptoms of dementia such as reduced working memory, information processing ability and speed, and a lack of ability to disregard unwanted information [31]. Therefore, traditional models of technology acceptability and robot-specific models may be inadequate to explain the acceptance factors pertaining to some OA, particularly those with dementia.

Models of Gerontechnology acceptance have been developed which do consider physical and cognitive health. Gerontechnology has been defined as electronic or digital products or services that can increase independent living and the social participation of OA [30]. Two examples of models are described below.

2.2.3 Acceptance models adapted for gerontechnology

The Senior Technology Acceptance Model (STAM) [30] was developed with data collected by personal interviews with community-dwelling people (n = 1012) over 55 years living in Hong Kong. STAM extends the constructs of previous technology acceptance models by adding age-related health and ability constructs: self-reported health conditions, cognitive abilities, attitudes towards ageing and life satisfaction, social relationships, and physical functioning. Chen and Chan [30] found that STAM can explain 68% of variance in gerontechnology usage and that facilitating conditions directly impacted the usage behaviour of participants. This suggested that the support of other people, knowledge, and guidance, directly impacts the gerontechnology usage of OA. The STAM has been used to appraise changes in technology acceptance in a randomised controlled trial that involved people with dementia (n = 103) living in residential care facilities (n = 7) [12]. The link between factors impacting the acceptance of a social robot, Kabochan, was assessed to ascertain if participant's attitudes and beliefs towards the robot would be impacted by the amount and the quality of engagement they had with the robot. They found that PEOU changed but beliefs and attitudes remained unchanged despite engagement with the robot.

The Model of Gerontechnology acceptance [32] encompasses both disability and aspects of a person's living environment. It was developed within the discipline of social gerontology, developed based on a study involving OA (n = 67) living in their own homes who were interviewed in-depth to ascertain their usage of and experience with a range of assistive technology (AT). AT was defined as being any device or system that allows an individual to perform a task that they would otherwise be unable to do. McCreadie and Tinker [32] argue that acceptability of AT is impacted by the interaction between a 'felt need' for assistance, and recognition of the product's quality regarding its efficiency, reliability, simplicity and safety, availability, and cost. They found four user attributes that were particularly relevant: OA disability, living arrangements, carer needs, and personal motivations and preferences. These attributes combined to generate a felt need for assistance from the technology, which combined with external factors that impacted access to the AT: information, contacts with suppliers, knowing what help is available. McCreadie and Tinker [32] emphasis that for AT to be acceptable, an OA must have information about how an AT might address their needs and be beneficial to them. In addition, OA acted in reciprocal relationships, with a strong desire for independence and to exercise their autonomy and agency in deciding whether to accept and AT [32]. The importance of autonomy and the agency of OA and people with dementia is discussed further below.

It is apparent that although gerontechnology models of acceptance do not address constructs that are particular to social robots as embodied social presences, they do highlight other factors that are pertinent to the way OA and people with dementia evaluate the acceptance of technology. Indeed, all the types of conceptual models currently available to guide empirical research in this context appear to have merits and limitations. It is also interesting to note that a recent systematic review, that included n = 74 studies, found that most studies that examine the technology acceptance, adoption, and the usability of information communication technology for people with dementia and their caregiver partners, do not report being underpinned by theoretical models and most use bespoke approaches to measuring acceptability [20]. Therefore, it is evident that a more consistent approach to examining [20] and conceptualizing robot acceptability for OA and people with dementia needs to be developed. It may be helpful for this approach to include constructs that are pertinent to robot usage and includes factors used in gerontechnological models.

The empirical research will now be examined to identify the factors that impact robot acceptability and what is important to OA and people with dementia and is likely to impact their decision whether to use a robot.

2.3 Factors that impact robot acceptance of people with dementia and older adults

OA and people with dementia are motivated to use robots based on factors that impact them at the level of individual people and other factors that relate to people that are known to individuals (significant others), and wider society. This section of the chapter will firstly discuss how individual-level factors impact the motivation of potential and actual users to accept robots. It then addresses significant others and cultural factors.

The empirical studies included in this was obtained through searching eight databases using the following search terms: accept*, dementia*, Alzheimer*, robot*, "cognitive deficiency", elder*, old*, technology accept*, user accept*, attitude, social robots, assistive technology. Papers were selected if they were published between 2005 and 2021, they involved people with dementia, and/or adults over 65 years old, if they were in English, and focused on robot acceptability.

2.3.1 Individual factors that impact the motivation of OA and people with dementia to accept social robots

Motivation to use robots is strongly related to their perceived usefulness. Social robots need to be relevant to the current unmet needs of potential users in order for robots to be perceived as useful [4, 14, 19, 28, 33–36]. Hebesberger et al. [36] aimed to investigate acceptance and the experience of using a humanoid robotic platform SCITOS, in a care institution in Australia. Data was collected in this 5-day pilot followed by a 15-day trial using semi-structured interviews and observations with people with dementia and their formal caregivers. They found that the robot was not perceived as useful and participants were ambivalent about the robot. Hebesberger et al. argues that acceptability is contingent on the robot meeting the specific needs of the end-users.

The identification of needs is complicated because OA and people with dementia may lack awareness of their unmet due to their cognitive difficulties [33], or because caregivers currently fulfil their needs [37] or because individuals are habituated to the challenges they are living with and don't perceive them as problems. But, people with dementia are very able to determine and articulate what they want and don't want and they know how they feel at any given moment [38, 39]. Potential users may also require information about the benefits of using the robot because they are new and unfamiliar technologies. Furthermore, the benefits of using the robot have to be real and clearly communicated to potential users [4, 28]. It is important to facilitate the development of realistic expectations to avoid users being disappointed if the expectations of using the robot are not met. This can result in a subsequent lack of acceptance [17].

In order to realise the benefit from the robot and to be motivated to use them, OA and people with dementia need to be capable of using a robot and to feel at ease doing so [16, 40]. This implies that there needs to be a user-technology fit [16] in terms of the robot capabilities to meet the users' needs. Thordardottir et al's., [17] systematic review that synthesised knowledge on the facilitators and barriers related to acceptance and the use of technology for people with cognitive impairment and caregivers found that PEOU was important and that acceptance was facilitated when technologies made low technical demands on users. A recent scoping review [41] also identified the robots need to be easy to use, to facilitate PEOU. This review included n = 53 studies and examined the use of barriers and facilitators affecting the implementation of zoomorphic robots by OA and people with dementia.

Ha and Park [42] in their survey found that increasing age may present OA with more obstacles to technology acceptance, including cognitive, motor, and sensory deficits. They also found that PEOU negatively correlated with support, i.e. there was a greater need for support when PEOU was low. But, PEOU can be changed through the support of skilled facilitators. In a study that examined the effect of the robot MARIO on the resilience of people with dementia (n = 10) [6], it was found that all the participants needed facilitation and then they successfully used MARIO's touchscreen and interacted with the robot. Chan et al. [12] also found PEOU is important and that it can be improved through engagement with a robot. This randomised controlled trial investigated how participants' attitudes and beliefs towards a humanoid robot, Kabochan that resembles a 3-year-old boy, would change, and whether a change was affected by the amount and quality of engagement with the robot. People with dementia (n = 103) living in residential care facilities (n = 7) were allocated into a group to engage with Kabochan or to a control group who received usual care. An ABAB withdrawal experimental design was used with each phase lasting 8 weeks. The controlled group received usual care throughout the 32-week study. The experimental condition received usual care during the first baseline phase and the third phase (A phases). In the second phase, the robot-engagement group had 2 weeks of introductory sessions with an occupational therapist and they became familiar with Kabochan's interactive features and how it could be used and turned on and off. Then for six weeks and during the 8 weeks of the 4th phase participants could use Kabochan as they wanted independently 24 hours a day. This study found that PEOU changed significantly ($p = -.042$) and that it was related to the intensity of constructive behavioural engagement that participants had with the robot.

PEOU is related to self-efficacy which is an important outcome measure of positive psychology in dementia that is related to a feeling of being in control [43]. In order for users to feel at ease, they need to feel in control whilst they are interacting with robots as autonomous embodied devices. In the study described above Korblet [18] found that self efficacy was mostly mentioned by the dementia group, and not by people with physical disabilities or visitors to the nursing home centre and that lack of self-efficacy led to participants not wanting to use the mobile presence robot again. Control requires a lack of fear, but also that the robot is reliable and can function properly to perform its tasks [10]. Conversely, technical robot problems and/or a lack of robot capability can threaten acceptance [36, 44]. Frennert et al. [45] asked OA with moderate sensory and mobility impairments to state their preference for an ideal robot and OA (n = 7) who lived with mock-ups of these ideals for one week. They found that feelings of control were also linked to feelings of trust and the need for privacy. Trust may be particularly important for OA who may mistrust technology more than younger adults [40]. This was determined in an online survey conducted in Japan, with respondents (n = 100; aged 20–70). Trust is also related to the degree that a robot is autonomous and adaptive in its movement and behaviour. Korblet [18] found that trust did not play a role for participants in their research when the robot was not autonomous and it was only used for a few minutes on one occasion in a group setting. Rossi et al. [9] also found that trust and anxiety did impact technology acceptance. The more autonomous and adaptable a robot is, the more challenging it is to human comfort, the ability to trust the robot, and the need for perceived control of the robot. It may be that when robots are used over a longer period, their requirement for adaptivity may increase as they are required to respond in a larger number of human social settings. De Graaf [46] examined the acceptability of Karotz, a rabbit-like robot, in the homes of OA (n = 6) over three 10 day periods. They found that as time passed, participants wanted more control of the robot to maintain their privacy because the robot continued to remind participants about their schedule and health-promoting activities in the presence of guests. Furthermore, how acceptable individual users find a robot depends on the individual person, the purpose of the robot, and the context [14].

The personality of the user, their interests, and values impact their response to a social robot and their perception about its social presence [9]. Rossi et al. found that having the personality trait of 'openness to experience' positively impacted the human-robot interaction. This study used a prototype of the humanoid robot Pepper to investigate if personality traits and user's empathy (a feature of personality) impacted the acceptance of a robot-led cognitive test. Participants were OA (n = 21; aged = 53–82 with average of 61). Acceptance of the robot was assess using, the UTAUT constructs and a psychologist evaluated personality traits, empathy, using the NEO Personality Inventory-3 (NEO-PI-3) [47]; the Empathy Quotient (EQ) [48]. After the psychological evaluation, the robot administered to participants psychometric tasks of the MoCA [49], and in a second task the person performed activities whilst Pepper monitored them. Dialogues of the participant-robot conversation and videos of the interaction were captured by Pepper. Rossi et al. found that empathy correlates with the amount the user perceived that the robot was sociable.

People with dementia can experience volatility in their mood and difficulty regulating their emotions [39]. Variations in mood are likely to impact robot acceptability [28]. When MARIO was being used by people with dementia, who had moderate and severe dementia, facilitators had to support participants to enable them to be ready to use the robot [6]. If participants were disorientated in time or space, facilitators had to acknowledge their perception of reality at that time and help the participants to deal with whatever was causing their anxiety. For example, at the start of one session with the robot, the participant welcomed MARIO and the facilitator into her room in the nursing home, but she was concerned, believing erroneously that a person had been into her house without her permission. This participant was not ready to use MARIO until she was calmed as a result of talking with the facilitator and being reassured that all was well [6].

OA and people with dementia are also motivated to use a robot if they find that interacting with it is enjoyable [19, 41, 50]. Novelty effects may enhance the enjoyment of robot usage initially, but these can decrease over time [46, 51]. How robots can be used to sustain a person's enjoyment and its relevance to their needs and capabilities, are discussed below. But first, the factors involving significant others and wider society that impact an individual's motivation to accept robots will be introduced.

2.3.2 Factors that involve significant others and wider society

The perceptions of significant others, particularly caregivers can have a substantial impact on robot acceptability. Significant others, as social influences, are strong predictors of the adoption of home healthcare robots [52] and the ITU the Kompai robot [53]. Significant others can enable the usage of robots through encouragement and facilitation [6, 53] or they can impede a robot's implementation into care settings [36, 44]. There is substantial evidence that professional health and social caregivers may have negative preconceptions about the use of robots for OA and people with dementia [44]. Caregivers may be concerned about the compatibility of robots with their work processes [41]. Casey et al. [8] explored the perceptions and experiences of using MARIO with people with dementia (n = 38), relatives/carers (n = 28), formal carers (n = 28) and managers (n = 13) in UK, Italy, and Ireland. They found that although MARIO was positively received by most of the participants, some formal care workers voiced concern that robots might replace care staff.

Human behaviour is impacted by a person's perceptions about what other people think, including thoughts related to negative ageing stereotypes that are prevalent in society [54]. People are motivated to act because they want to project a self-image that they are healthy and independent, both to themselves and other people [33, 55, 56]. The acceptance of robots by people with dementia and OA is impacted as potential users may fear being stigmatised and labelled as physically or psychologically vulnerable, dependent, in decline, or lonely if they use a robot [34, 44, 53, 56]. Dudek et al. introduced OA (n = 28) in good subjective physical and cognitive health to PLEO, a robot shaped like a dinosaur, and then assessed robot acceptance using an Emotional Attachment Scale (EA-Scale) developed by Thomson et al. [57] and the Comfort from Companion Animals Scale developed by Zasloff [58]. Then they accessed the participants' actual and ideal self and subjective robot user image. They found that participants stigmatised OA robot users and that OA may construct a more negative user image than their own self-image, in order to maintain their own positive self-image. Dudek et al. concluded that acceptance of new technologies is impacted by OA subjective interpretations of the technologies.

Another factor that impacts the level of robot acceptability for individuals, significant others, and wider society, is the familiarity of robots as a technology [44]. It is argued that people need time to learn about and become comfortable with robots as a technology [17, 44, 51] and that this may happen when robots are more prevalent. The level of robot acceptability will vary between cultures and over time, but as robots become more available and diffuse in society it is not certain that acceptability will increase [28]. Bishop et al. found that higher robot familiarity had a negative impact on attitudes and behaviours and that familiarity was linked to heightened awareness of a robot's limitations. How robots need to be design, developed, and deployed used in order to facilitate their acceptability will now be discussed.

2.4 What is likely to increase robot acceptability for OA and people with dementia

Robot acceptance is increased through personalisation of the robot, to make it meet the needs and preferences of the user and support their personhood [7, 14, 17, 44]. Personhood has been defined as 'a standing or a status that is bestowed on one human being by another, in the context of relationship and social being' [59] p.8. For example, the acceptability of the robot Maltida was increased due to personalisation that supported personhood, through the robot using human-like emotive expressions and accounting for the user's disabilities [7]. Other ways to support personhood and increase the acceptability of a robot is if the robot enables the OA and person with dementia to feel empowered, respected, and able to participate in activities that are meaningful to them [6, 10].

It is well recognised in current discourses about dementia care, that people with dementia do not experience a 'loss of self' as dementia progresses [39]. Whilst a person's identity, and personality may change, a person's central being, core values remain [39] and people with dementia maintain the potential to adapt and grow [38]. It is not enough to discern the requirements of the robot user on one occasion and personalise the robot. To ensure ongoing acceptance of the robot, there must be ongoing review and adaptation of the robot as the person uses it over time. The investigation with MARIO described above found that the creation of meaningful activity was made possible through a skilled facilitator and person with dementia working in partnership to use the robot in a way that was meaningful to the user [6]. Facilitators learned about the user's preferences and used the robot creatively in ways that were not identified before usage, even though MARIO was initially personalised for each individual using information obtained from each person with dementia, their caregivers, and relatives. For example, the following conversation was observed between a participant, Margaret, who had moderate dementia, and a facilitator whilst they used MARIO's applications. It illustrates how new knowledge of a participant's interests were used to further personalise MARIO.

The facilitator and Margaret were chatting through the photographs then Margaret chose the music application.

> *Margaret 'It's very good', looking at MARIO's face while the music is playing for 1 minute and then she says, 'I would like to get home'.*
>
> *Facilitator 'Yes. Does the music remind you of something?'*
>
> *Margaret 'I would like to do that myself ... .. the same as other people'*
>
> *Facilitator 'You'd like to be more independent?'*
>
> *Margaret 'Yes (pause) ... Do you like the music?'*
>
> *Facilitator 'Yes, it's lovely ... ..does it remind you of something?'*
>
> *Margaret 'Jeanie of the light brown hair'.*
>
> *Facilitator 'Is that a song?'*
>
> *Margaret 'Yes'*
>
> *Facilitator 'Would you like MARIO to play it?'*
>
> *(OME Margaret Session 10, [6]).*

Another way in which robots may be made more acceptable is if their investigation, design, development, and deployment are guided by models of gerontology that focus on the strengths and abilities of OA and people with dementia, rather than their disabilities. New ways of using robots may be possible if the lens of successful aging or resilience as strength-based models of care are applied to social robots [6, 44]. Such models would facilitate the development of guidelines and protocols for robot usage that aim to ensure the autonomy and dignity of OA and people with dementia are upheld [12] and that their priorities and goals are central to the robot design and usage.

The robot's physical appearance, behaviour, and communication style, and ability, combine to impact the robot's social presence and perceived sociability. There is huge variability in the optimal appearance of social robots [14]. But, robot acceptability will increase when robots are technically improved and able to behave and communicate in a convincing human-like way [14]. To support human communication, robots need to: enable a convincing emotional exchange; understand the user's intention; and provide complementary reactions [28]. This is because people anthropomorphise about a robot [60] and want this autonomous embodied presence to be compatible with human norms of behaviour, (or that of an animal if it is a zoomorphic robot). There is some evidence that OA are more likely to regard robot interaction as pleasurable if robots are designed to display positive emotion [28]. However, the user must interpret that the robot behaves in a way that is compatible with their status and is appropriate for the psych-social context [61, 62].

Cobot or collaborative robots are currently being developed and used in the manufacturing industry. These robots are designed to move and perform delegated repetitive tasks, working alongside humans. Unlike other robots, Cobots can be taught by human users through example, rather than through programming. As such they are intended for direct human-robot interaction, to be used in a shared space in close proximity to humans. They can be smaller, more mobile, and they are reputedly safer than their traditional robotic counterparts [63]. No research to date has explored the usage of collaborative robots in health and social care settings. However, it may be possible to design collaborative robots that support the independence of people with dementia and older adults and that support the work of human caregivers. As discussed above, one of the moral and ethical objections to using robots in the health and social care setting is that human-human contact will be reduced. Because collaborative robots work alongside human workers, as a tool, these fears may be mitigated. In addition, the ability to learn by example could make collaborative robots easier to use in a clinical and practical context by people who don't have programming capability. Collaborative robots have the potential to be used more flexibly in response to the individual needs of users and to meet their requirements as these needs change. It has been argued above that flexibility, attention to individual needs, and ease of use, are key requirements for robots to be acceptable to users in the health and social care context. Collaborative robots could potentially help individuals, alongside the care provided by humans, to take food and drink and to perform other daily living activities. Indeed, the collaborative element of cobots could also improve the person-centred usability of other robotic and non-robotic devices that currently assist the movement of people who are physically impaired.

The acceptability of all robots will be enhanced if robots are designed, developed, and deployed with the significant, early, and repeated input of OA, people with dementia, and significant others, including health and social care professionals. All these stakeholders need to work in partnership with researchers and developers, in participatory design processes [11] so that robots are optimally relevant and able to meet the needs of users. Iterative participatory design was used in the development of the robots Matilda and MARIO. Both robots were improved and customised to meet the needs of people with dementia with prolonged, iterative phases that involved the assessment of needs [64], development using the feedback of users, and testing in clinical environments [7, 8].

There have been some developments that will facilitate the participation of stakeholders in research and will encourage a focus on issues of safety and ethics [44]. A European Commission framework requires that users, innovators, and society mutually interact and engage in processes that are transparent so that products are developed to be acceptable, sustainable, and desirable [65]. Dementia-specific developments have also been advanced. For example, in Europe, the Alzheimer's Society has produced comprehensive guidelines on how the ethical challenges of involving people with dementia in research can be managed [66] and best practice guidelines have been produced [67]. In addition, a European funded project (prospero.via.dk/en) is also underway that aims to understand what health and social care professional needs regarding robotic technologies and how caregivers can contribute to the development of robots.

Very few longitudinal studies have investigated robot acceptability [14, 20]. Robot acceptability will also improve if developers and researchers consider, in-depth, the context into which the robot is going to be deployed. The technology must be examined in the context it will be used, with longitudinal research designs [44], for over two months [4]. An in- depth understanding of the context, including the needs and motivations of all stakeholders, will also be facilitated by triangulating the data from multiple data sources that include qualitative research methods [4, 6]. Currently, robot acceptance is mostly measured using indirect methods such as questionnaires, and interviews, rather than direct observation during human-robot interactions [20]. Observational 'in the moment' methods are particularly advantageous to understand the experiences of people with dementia, who may not recall accurately and in detail, their experiences after interactions have taken place. Furthermore, people with severe dementia may be unable to use questionnaires, and reliance on proxy recordings of their attitude and beliefs [12] may not be accurate.

3. Conclusion

The chapter has defined the acceptability of social robots and discussed how acceptability can be understood and predicted as conceptualized by theoretical models of acceptability. It has discussed factors that impact the acceptability of social robots by OA and people with dementia and made recommendations as to how acceptability may be increased. It has been argued that current acceptance models for social robots need further development to accommodate the needs of OA and people with dementia who experience physical and cognitive disability. Models of gerontechnology may be useful to identify the needs and characteristics of OA and people with dementia that are pertinent to robot acceptability. Models need further development and testing to ensure they can fully inform both the acceptability evaluations of autonomous embodied social robots.

The multiple interacting factors that impact robot acceptance at the individual, community and societal levels have been discussed. It is vitally important that robots are useful, enjoyable to use, easy to use and that they support the personhood and strengths of OA and people with dementia. Enjoyable social interaction requires robots to have good quality humanlike communication skills. It would enhance the acceptability of robots if future design, development, and deployment of robots are underpinned by strength-based theories. This would ensure that processes are driven by the motivations and goals of OA and people with dementia. It is also important that the context surrounding robot deployment is assessed in-depth using longitudinal designs and that all aspects of their development are guided by potential users, significant others and health, and social care stakeholders.

Acknowledgements

Funding: The research leading to these results has received funding from the European Union Horizons 2020-the Framework Programme for Research and Innovation (2014-2020) under grant agreement 643808 Project MARIO 'Managing active and healthy aging with use of caring service robots'.

Conflict of interest

The authors declare that they have no conflict of interest.

Author details

Sally Whelan* and Dympna Casey
National University of Ireland Galway, Ireland

*Address all correspondence to: sally.whelan@nuigalway.ie

References

[1] Wimo, A., and Prince, M.J.: 'World Alzheimer Report 2010: the global economic impact of dementia' (Alzheimer's Disease International, 2010. 2010)

[2] Prince, M., Guerchet, M., and Prina, M.: 'Policy Brief for Heads of Government: The Global Impact of Dementia 2013-2050', in Editor (Ed.)^(Eds.): 'Book Policy Brief for Heads of Government: The Global Impact of Dementia 2013-2050' (Alzheimer's disease International (ADI), 2013, edn.), pp.

[3] Broekens, J., Heerink, M., and Rosendal, H.: 'Assistive social robots in elderly care: a review', Gerontechnology, 2009, 8, (2), pp. 94-103

[4] de Graaf, M.M.A.: 'Living with robots: investigating the user acceptance of social robots in domestic environments', University of Twente, Enschede, 2015

[5] Commission, E.: 'Long-term care in ageing societies–Challenges and policy options', in Editor (Ed.)^(Eds.): 'Book Long-term care in ageing societies–Challenges and policy options' (Belgium: Brussel, 2013, edn.), pp.

[6] Whelan, S., Burke, M., Barrett, E., Mannion, A., Kovačič, T., Santorelli, A., Luz Oliveira, B., Gannon, L., Shiel, E., and Casey, D.: 'The effects of MARIO, a social robot, on the resilience of people with dementia: A multiple case study', Gerontechnology, 2020.

[7] Khosla, R., Nguyen, K., and Chu, M.-T.: 'Human robot engagement and acceptability in residential aged care', International Journal of Human–Computer Interaction, 2017, 33, (6), pp. 510-522

[8] Casey, D., Barrett, E., Kovacic, T., Sancarlo, D., Ricciardi, F., Murphy, K., Koumpis, A., Santorelli, A., Gallagher, N., and Whelan, S.: 'The Perceptions of People with Dementia and Key Stakeholders Regarding the Use and Impact of the Social Robot MARIO', International Journal of Environmental Research and Public Health, 2020, 17, (22), pp. 8621

[9] Rossi, S., Conti, D., Garramone, F., Santangelo, G., Staffa, M., Varrasi, S., and Di Nuovo, A.: 'The role of personality factors and empathy in the acceptance and performance of a social robot for psychometric evaluations', Robotics, 2020, 9, (2), pp. 39.

[10] Scopelliti, M., Giuliani, M.V., and Fornara, F.: 'Robots in a domestic setting: a psychological approach', Universal access in the information society, 2005, 4, (2), pp. 146-155.

[11] Ozdemir, D., Cibulka, J., Stepankova, O., and Holmerova, I.: 'Design and implementation framework of social assistive robotics for people with dementia-a scoping review', Health and Technology, 2021, pp. 1-12.

[12] Chen, K., and Lou, V.W.Q.: 'Measuring senior technology acceptance: development of a brief, 14-item scale', Innovation in aging, 2020, 4, (3), pp. igaa016.

[13] Broadbent, E., Tamagawa, R., Kerse, N., Knock, B., Patience, A., and MacDonald, B.: 'Retirement home staff and residents' preferences for healthcare robots', in Editor (Ed.)^(Eds.): 'Book Retirement home staff and residents' preferences for healthcare robots' (IEEE, 2009, edn.), pp. 645-650.

[14] Whelan, S., Murphy, K., Barrett, E., Krusche, C., Santorelli, A., and Casey, D.: 'Factors affecting the acceptability of social robots by older adults including people with dementia or cognitive impairment: a literature review',

International Journal of Social Robotics, 2018, 10, (5), pp. 643-668

[15] De Graaf, M.M., and Allouch, S.B.: 'Exploring influencing variables for the acceptance of social robots', Robotics and autonomous systems, 2013, 61, (12), pp. 1476-1486.

[16] Baisch, S., Kolling, T., Schall, A., Rühl, S., Selic, S., Kim, Z., Rossberg, H., Klein, B., Pantel, J., and Oswald, F.: 'Acceptance of social robots by elder people: does psychosocial functioning matter?', International Journal of Social Robotics, 2017, 9, (2), pp. 293-307.

[17] Thordardottir, B., Malmgren Fänge, A., Lethin, C., Rodriguez Gatta, D., and Chiatti, C.: 'Acceptance and use of innovative assistive technologies among people with cognitive impairment and their caregivers: a systematic review', BioMed research international, 2019.

[18] Korblet, V.: 'The acceptance of mobile telepresence robots by elderly people', University of Twente, 2019

[19] Heerink, M., Kröse, B., Evers, V., and Wielinga, B.: 'Assessing acceptance of assistive social agent technology by older adults: the almere model', International journal of social robotics, 2010, 2, (4), pp. 361-375.

[20] Cruz, M.A., Daum, C., Comeau, A., Salamanca, J.D.G., McLennan, L., Neubauer, N., and Liu, L.: 'Acceptance, adoption, and usability of information and communication technologies for people living with dementia and their care partners: a systematic review', Disability and Rehabilitation: Assistive Technology, 2020, pp. 1-15.

[21] Peters, M., Schneider, M., Griesshaber, T., and Hoffmann, V.H.: 'The impact of technology-push and demand-pull policies on technical change–Does the locus of policies matter?', Research Policy, 2012, 41, (8), pp. 1296-1308.

[22] Tapus, A., Tapus, C., and Mataric, M.: 'The role of physical embodiment of a therapist robot for individuals with cognitive impairments', in Editor (Ed.)^(Eds.): 'Book The role of physical embodiment of a therapist robot for individuals with cognitive impairments' (IEEE, 2009, edn.), pp. 103-107

[23] Ajzen, I.: 'The theory of planned behavior. Orgnizational Behavior and Human Decision Processes, 50, 179-211', in Editor (Ed.)^(Eds.): 'Book The theory of planned behavior. Orgnizational Behavior and Human Decision Processes, 50, 179-211' (1991, edn.), pp.

[24] Stafford, R., Broadbent, E., Jayawardena, C., Unger, U., Kuo, I.H., Igic, A., Wong, R., Kerse, N., Watson, C., and MacDonald, B.A.: 'Improved robot attitudes and emotions at a retirement home after meeting a robot', in Editor (Ed.)^(Eds.): 'Book Improved robot attitudes and emotions at a retirement home after meeting a robot' (IEEE, 2010, edn.), pp. 82-87

[25] Davis, F.D.: 'Perceived usefulness, perceived ease of use, and user acceptance of information technology', MIS quarterly, 1989, pp. 319-340.

[26] Aldhaban, F.: 'Exploring the adoption of Smartphone technology: Literature review', in Editor (Ed.)^(Eds.): 'Book Exploring the adoption of Smartphone technology: Literature review' (IEEE, 2012, edn.), pp. 2758-2770

[27] Venkatesh, V., Morris, M.G., Davis, G.B., and Davis, F.D.: 'User acceptance of information technology: Toward a unified view', MIS quarterly, 2003, pp. 425-478.

[28] Bishop, L., van Maris, A., Dogramadzi, S., and Zook, N.: 'Social robots: The influence of human and robot characteristics on acceptance', Paladyn, Journal of Behavioral Robotics, 2019, 10, (1), pp. 346-358.

[29] Heerink, M.: 'Exploring the influence of age, gender, education and computer experience on robot acceptance by older adults', in Editor (Ed.)^(Eds.): 'Book Exploring the influence of age, gender, education and computer experience on robot acceptance by older adults' (IEEE, 2011, edn.), pp. 147-148

[30] Chen, K., and Chan, A.H.S.: 'Gerontechnology acceptance by elderly Hong Kong Chinese: a senior technology acceptance model (STAM)', Ergonomics, 2014, 57, (5), pp. 635-652.

[31] Tenneti, R., Johnson, D., Goldenberg, L., Parker, R.A., and Huppert, F.A.: 'Towards a capabilities database to inform inclusive design: Experimental investigation of effective survey-based predictors of human-product interaction', Applied ergonomics, 2012, 43, (4), pp. 713-726.

[32] McCreadie, C., and Tinker, A.: 'The acceptability of assistive technology to older people', Ageing and society, 2005, 25, (1), pp. 91-110.

[33] Stafford, R.Q., MacDonald, B.A., and Broadbent, E.: 'Identifying specific reasons behind unmet needs may inform more specific eldercare robot design', in Editor (Ed.)^(Eds.): 'Book Identifying specific reasons behind unmet needs may inform more specific eldercare robot design' (Springer, 2012, edn.), pp. 148-157

[34] Pino, M., Boulay, M., Jouen, F., and Rigaud, A.S.: '"Are we ready for robots that care for us?" Attitudes and opinions of older adults toward socially assistive robots', Frontiers in aging neuroscience, 2015, 7, pp. 141.

[35] Mitzner, T.L., Chen, T.L., Kemp, C.C., and Rogers, W.A.: 'Identifying the potential for robotics to assist older adults in different living environments', International journal of social robotics, 2014, 6, (2), pp. 213-227.

[36] Hebesberger, D., Koertner, T., Gisinger, C., and Pripfl, J.: 'A long-term autonomous robot at a care hospital: A mixed methods study on social acceptance and experiences of staff and older adults', International Journal of Social Robotics, 2017, 9, (3), pp. 417-429.

[37] Begum, M., Wang, R., Huq, R., and Mihailidis, A.: 'Performance of daily activities by older adults with dementia: The role of an assistive robot', in Editor (Ed.)^(Eds.): 'Book Performance of daily activities by older adults with dementia: The role of an assistive robot' (IEEE, 2013, edn.), pp. 1-8

[38] Sabat, S.R.: 'Alzheimer's Disease and Dementia: What Everyone Needs to Know®' (Oxford University Press, 2018. 2018)

[39] Bryden, C.: 'Will I Still be Me?: Finding a Continuing Sense of Self in the Lived Experience of Dementia', Jessica Kingsley Publishers, 2018.

[40] Nomura, T., Sugimoto, K., Syrdal, D.S., and Dautenhahn, K.: 'Social acceptance of humanoid robots in Japan: A survey for development of the frankenstein syndorome questionnaire', in Editor (Ed.)^(Eds.): 'Book Social acceptance of humanoid robots in Japan: A survey for development of the frankenstein syndorome questionnaire' (IEEE, 2012, edn.), pp. 242-247

[41] Koh, Q.W., Feling, S.A., Budak, B., Toomey, E., and Casey, D.: 'Barriers and facilitators to the implementation of social robots for older adults and people with dementia: A scoping review. ', Submitted to BMC Geriatrics, (in review)

[42] Ha, J., and Park, H.K.: 'Factors Affecting the Acceptability of Technology in Health Care Among Older Korean Adults with Multiple Chronic Conditions: A Cross-Sectional Study Adopting the Senior Technology

Acceptance Model', Clinical Interventions in Aging, 2020, 15, pp. 1873.

[43] Stoner, C.R., Stansfeld, J., Orrell, M., and Spector, A.: 'The development of positive psychology outcome measures and their uses in dementia research: a systematic review', Dementia, 2019, 18, (6), pp. 2085-2106.

[44] Papadopoulos, I., Koulouglioti, C., Lazzarino, R., and Ali, S.: 'Enablers and barriers to the implementation of socially assistive humanoid robots in health and social care: a systematic review', BMJ open, 2020, 10, (1), pp. e033096.

[45] Frennert, S., Eftring, H., and Östlund, B.: 'Older people's involvement in the development of a social assistive robot', in Editor (Ed.)^(Eds.): 'Book Older people's involvement in the development of a social assistive robot' (Springer, 2013, edn.), pp. 8-18

[46] De Graaf, M.M., Allouch, S.B., and Klamer, T.: 'Sharing a life with Harvey: Exploring the acceptance of and relationship-building with a social robot', Computers in human behavior, 2015, 43, pp. 1-14.

[47] McCrae, R.R., Costa, J., Paul T, and Martin, T.A.: 'The NEO–PI–3: A more readable revised NEO personality inventory', Journal of personality assessment, 2005, 84, (3), pp. 261-270.

[48] Baron-Cohen, S., and Wheelwright, S.: 'The empathy quotient: an investigation of adults with Asperger syndrome or high functioning autism, and normal sex differences', Journal of autism and developmental disorders, 2004, 34, (2), pp. 163-175.

[49] Nasreddine, Z.S., Phillips, N.A., Bédirian, V., Charbonneau, S., Whitehead, V., Collin, I., Cummings, J.L., and Chertkow, H.: 'The Montreal Cognitive Assessment, MoCA: a brief screening tool for mild cognitive impairment', Journal of the American Geriatrics Society, 2005, 53, (4), pp. 695-699.

[50] Young, J.E., Hawkins, R., Sharlin, E., and Igarashi, T.: 'Toward acceptable domestic robots: Applying insights from social psychology', International Journal of Social Robotics, 2009, 1, (1), pp. 95-108.

[51] Torta, E., Werner, F., Johnson, D.O., Juola, J.F., Cuijpers, R.H., Bazzani, M., Oberzaucher, J., Lemberger, J., Lewy, H., and Bregman, J.: 'Evaluation of a small socially-assistive humanoid robot in intelligent homes for the care of the elderly', Journal of Intelligent & Robotic Systems, 2014, 76, (1), pp. 57-71.

[52] Alaiad, A., and Zhou, L.: 'The determinants of home healthcare robots adoption: An empirical investigation', International journal of medical informatics, 2014, 83, (11), pp. 825-840.

[53] Wu, Y.-H., Wrobel, J., Cornuet, M., Kerhervé, H., Damnée, S., and Rigaud, A.-S.: 'Acceptance of an assistive robot in older adults: a mixed-method study of human–robot interaction over a 1-month period in the Living Lab setting', Clinical interventions in aging, 2014, 9, pp. 801.

[54] Dudek, M., Baisch, S., Knopf, M., and Kolling, T.: '"THIS ISN'T ME!": The Role of Age-Related Self-and User Images for Robot Acceptance by Elders', International Journal of Social Robotics, 2020, pp. 1-15.

[55] Frennert, S., and Östlund, B.: 'Seven matters of concern of social robots and older people', International Journal of Social Robotics, 2014, 6, (2), pp. 299-310.

[56] Neven, L.: ''But obviously not for me': robots, laboratories and the defiant identity of elder test users', Sociology of health & illness, 2010, 32, (2), pp. 335-347.

[57] Thomson, M., MacInnis, D.J., and Park, C.W.: 'The ties that bind: Measuring the strength of consumers' emotional attachments to brands', Journal of consumer psychology, 2005, 15, (1), pp. 77-91.

[58] Zasloff, R.L.: 'Measuring attachment to companion animals: a dog is not a cat is not a bird', Applied Animal Behaviour Science, 1996, 47, (1-2), pp. 43-48.

[59] Kitwood, T., and Bredin, K.: 'Towards a theory of dementia care: personhood and well-being', Ageing & Society, 1992, 12, (3), pp. 269-287.

[60] Stafford, R.Q., MacDonald, B.A., Jayawardena, C., Wegner, D.M., and Broadbent, E.: 'Does the robot have a mind? Mind perception and attitudes towards robots predict use of an eldercare robot', International journal of social robotics, 2014, 6, (1), pp. 17-32.

[61] Flandorfer, P.: 'Population ageing and socially assistive robots for elderly persons: the importance of sociodemographic factors for user acceptance', International Journal of Population Research, 2012.

[62] Heerink, M., Kröse, B., Wielinga, B., and Evers, V.: 'Enjoyment intention to use and actual use of a conversational robot by elderly people', in Editor (Ed.)^(Eds.): 'Book Enjoyment intention to use and actual use of a conversational robot by elderly people' (2008, edn.), pp. 113-120

[63] Casey, D., Felzmann, H., Pegman, G., Kouroupetroglou, C., Murphy, K., Koumpis, A., and Whelan, S.: 'What people with dementia want: designing MARIO an acceptable robot companion', in Editor (Ed.)^(Eds.): 'Book What people with dementia want: designing MARIO an acceptable robot companion' (Springer, 2016, edn.), pp. 318-325.

[64] Stahl, B.C., and Coeckelbergh, M.: 'Ethics of healthcare robotics: Towards responsible research and innovation', Robotics and Autonomous Systems, 2016, 86, pp. 152-161

[65] Watchman, K.: 'Overcoming ethical challenges affecting the involvement of people with dementia in research: recognising diversity and promoting inclusive research', Alzheimer Europe, 2020.

[66] Rai, H.K., Barroso, A.C., Yates, L., Schneider, J., and Orrell, M.: 'Involvement of people with dementia in the development of technology-based interventions: Narrative synthesis review and best practice guidelines', Journal of medical Internet research, 2020, 22, (12), pp. e17531.

[67] Share, P., and Pender, J.: 'Helping tomorrow's social professionals to learn about social robotics', 6th International Conference on Higher Education Advances (HEAd'20) Universitat Politècnica de València, València, 2020.

Chapter 3

Self-Learning Low-Level Controllers

Dang Xuan Ba and Joonbum Bae

Abstract

Humanoid robots are complicated systems both in hardware and software designs. Furthermore, the robots normally work in unstructured environments at which unpredictable disturbances could degrade control performances of whole systems. As a result, simple yet effective controllers are favorite employed in low-level layers. Gain-learning algorithms applied to conventional control frameworks, such as Proportional-Integral-Derivative, Sliding-mode, and Backstepping controllers, could be reasonable solutions. The adaptation ability integrated is adopted to automatically tune proper control gains subject to the optimal control criterion both in transient and steady-state phases. The learning rules could be realized by using analytical nonlinear functions. Their effectiveness and feasibility are carefully discussed by theoretical proofs and experimental discussion.

Keywords: backstepping control, PID control, sliding mode control, gain-learning control, position control, low-level control

1. Introduction

Precise motion control of low-level systems is one of the most important tasks in industrial and humanoid robotic systems [1–3]. Different from industrial robots which commonly operate in compact regions with simple and almost repetitive missions, humanoid robots perform complicated works and face to unknown disturbances in daily activities. Hence, designing a high-performance controller that is easy to use in real-time implementation for such the maneuver systems is a big challenge [4, 5].

To accomplish motion control in real-time applications, conventional proportional-integral-derivative (PID) controllers are the first selection from engineers and researchers thanks to simplicity in design and acceptable control outcome for uncertain systems [6–11]. Stability of the servo-controlled systems is proven by theoretical analyses, and their flexibility could be enhanced using machine-learning methods such as ordinary or neuro fuzzy-logic-based self-tuning [7, 10, 11], pole-placement adaptation [8], or convolutional learning [9]. However, using linear control signals to suppress the nonlinear behaviors of the robotic dynamics may lead to unexpected transient performance. To overcome this drawback, nonlinear controllers such as sliding mode control (SMC), backstepping control (BSC), or inverse dynamical control have gotten attention from developers [12–17]. Indeed, a robust-integral--sign-error (RISE) controller was studied to consolidate lumped disturbances inside the system dynamics for achieving asymptotic control results [13]. In another direction, a model-based nonlinear disturbance-observer controller was proposed based on the backstepping technique to yield excellent control accuracies [15]. Nevertheless, extended studies noted that the outstanding control performances are difficult to be preserved with hard control gains employed in diverse real-time operations [18, 19].

As a result, gain-learning SMC algorithms have been developed for robotic systems [18–21]. The control objective could be minimized by learning processes of robust gains, driving gains or massive gains [22, 23]. In fact, some control gains still need to manually tune for their possibly wide ranges due to nature of each control plant. Thus, it may lead to inconvenience during the operation.

Intelligent methods for automatically tuning all the control gains have been also proposed based on modified backtracking search algorithms (MBSA) combining with a Type-2 fuzzy-logic design [24] or model predictive approaches [25]. The desired gains could be estimated for the best performance by dealing with closed-loop optimal constraints. Though promising control results were presented, smooth variation of the gain dynamics need to further consideration.

Gain-learning control approaches under backstepping design provided another interesting direction as well. PID control with a gain-varying technique encoded by the backstepping scheme was formerly studied [26]. Success of the creative control method was confirmed by a thorough theoretical proof and experimental validation results. Since the learning process of all the control gain is generated only by one damping function, versatility of the control design may be limited for diverse working conditions. Improvement on the flexibility of gain selection is thus still an open issue.

In this chapter, an extensive gain-adaptive nonlinear control approach is presented for high-performance motion control of a low-level servo system. The controller is comprised of an inner robust nonlinear loop and an outer gain-learning loop. The inner loop is developed based on a RISE-modified backstepping framework to ensure asymptotic tracking control in the existence of nonlinear uncertainties and disturbances. The second loop contains a new gain-adaptive engine to activate variation gains of the inner loop in real-time applications. Theoretical effectiveness of the proposed controller is concretely proven by Lyapunov-based analyses. Feasibility of the control approach was confirmed by intensive real-time experiments on a legged robot. Their features are presented in detail in the below sections.

2. Problem statements

General dynamics of a robotic system could be expressed in the following form:

$$\mathbf{M}(\mathbf{q})\ddot{\mathbf{q}} + \mathbf{C}(\mathbf{q}, \dot{\mathbf{q}})\dot{\mathbf{q}} + \mathbf{g}(\mathbf{q}) + \boldsymbol{\tau}_{fr}(\dot{\mathbf{q}}) + \mathbf{J}^T\mathbf{f}_{ext} = \boldsymbol{\tau} \tag{1}$$

where $\mathbf{q}, \dot{\mathbf{q}}, \ddot{\mathbf{q}} \in \Re^n$ are respectively the joint position, velocity and acceleration vectors, $\mathbf{M}(\mathbf{q}) \in \Re^{n \times n}$is the inertia matrix, $\mathbf{C}(\mathbf{q}, \dot{\mathbf{q}}) \in \Re^{n \times n}$is the Centrifugal/Coriolis matrix, $\mathbf{g}(\mathbf{q}) \in \Re^n$denotes the gravitational torque, $\boldsymbol{\tau}_{fr}(\dot{\mathbf{q}}) \in \Re^n$ is the frictional torque, $\mathbf{J}^T$ is the respective Jacobian matrix, $\mathbf{f}_{ext}$ is the external disturbance, and $\boldsymbol{\tau}$ is control torque at robot joints.

The main control objective here is to find out a proper control signal $\boldsymbol{\tau}$ that ensures a control error between the system output and a desired profile stabilizing at origin under various complicated environments.

To realize the control objective, conventional linear or nonlinear controllers such as Proportional-Integral-Derivative (PID) and Sliding mode control (SMC) methods are priority selections in industry thanks to their simplicity and robustness. However, such the mission in humanoid robots is a different story in which the systems frequently operate in unknown environments with harshly unpredictable disturbances [27, 28]. Obviously, the required controller is strong robustness, fast adaptation, and easy implementation.

3. Low-level intelligent nonlinear controller

In this subsection, a position controller is developed based on the general model using the backstepping technique and new adaptation laws. The dynamics Eq. (1) can be splitted for low-level subsystems under the following state-space form:

$$\begin{cases} \dot{x}_1 = x_2 + \upsilon \\ \dot{x}_2 = -a_1 x_2 + a_2 u + d \end{cases} \tag{2}$$

where $x_1 = \mathbf{q}_{i|i=1..n}$ presents a specific joint angle, x_2 is the measurement joint velocity, $u = \boldsymbol{\tau}_{i|i=1..n}$ is the control torque at the specific joint, υ is the measurement noise, a_1 is a positive constant presenting the nominal dynamics, a_2 is another positive constant standing for the inverse nominal mass at low-level dynamics, and d is the lumped disturbance denoting the deviation of internal dynamics. Note that, x_1 and x_2 hold for the following assumptions:

Assumption 1:

a. The system output x_1 is measurable.

b. The angular velocity (x_2) is bounded and is indirectly measured from the angular data with a bounded tolerance (υ).

3.1 Robust backstepping control scheme

Let formulate the main control error as:

$$e_1 = x_1 - x_{1d} \tag{3}$$

where x_{1d} is the desired trajectory of the controlled joint.

Before designing the final control signal, additional assumptions are given.

Assumption 2:

a. The measurement noise υ is bounded and differentiable up to the second order.

b. The disturbance d and its time derivative are bounded.

c. The desired signal x_{1d} is bounded and differentiable up to the third order.

The time derivative of the control objective e_1 in considering the first equation of dynamics Eq. (2) is:

$$\dot{e}_1 = x_2 + \upsilon - \dot{x}_{1d} \tag{4}$$

To control the error e_1 to zero or to be as small as possible, a virtual control signal is employed to remove the time derivative of the desired signal and to compensate for the disturbance υ:

$$x_{2d} = \dot{x}_{1d} - k_1 e_1 \tag{5}$$

where k_1 is a positive constant.

A new state control error is defined as:

$$e_2 = x_2 - x_{2d} \tag{6}$$

Differentiating the new error with respect to time and using the second equation of the dynamics Eq. (2) lead to

$$\dot{e}_2 = -a_1 x_2 + a_2 u + d - \ddot{x}_{1d} + k_1(x_2 + \upsilon - \dot{x}_{1d}) \tag{7}$$

To drive the new control error e_2 to an expected range, the final control signal is proposed as follows, including two sub-control terms (a model-based term and robust term):

$$u = -a_2^{-1}\left(-a_1 x_2 + k_1(x_2 - \dot{x}_{1d}) + (k_2 - k_1)e_2 + (k_3 + k_4)e_1 + \int_0^t (k_4 k_1 e_1 + k_5 \operatorname{sgn}(e_1))d\tau\right) \tag{8}$$

where $k_{i|i=2,3,4,5}$ are positive control gains.

Stability of the closed-loop system under the controller Eq. (8) can be confirmed by the following statement.

Lemma 1:

Given a low-level system Eq. (2) under *Assumptions 1* and *2*, if employing the control rule Eqs. (3)–(8), stability of the closed-loop system is ensured for the positive bounded control gains $k_{i|i=1..5}$ satisfying:

$$k_2 - k_1 > 0 \tag{9}$$

Proof of *Lemma 1* is given in *Appendix A*.

Remark 1: Lemma 1 reveals that the closed-loop system is stabilized at a vicinity around zero under the constrain Eq. (9). Obviously, acceptable control performance could be resulted in with proper control gains selected.

Effectiveness of the nonlinear control structure is achieved by the following statement:

Theorem 1:

Given a closed loop system satisfying *Lemma 1*, it asymptotically converges if properly further choosing the control gains such that:

$$k_5 \geq \Delta_{\dot{h}} \tag{10}$$

Proof of *Theorem 1* is discussed in *Appendix B*.

Remark 2: In real-time situations [15, 29, 30], the position data x_1 are employed to approximate the velocity x_2 throughout a low-pass filter. Thus, the perturbance term (υ) obviously exists in the studied model Eq. (2) and its variation depends on the used filter.

Remark 3: With the robust backstepping control scheme designed, an excellent control performance can be resulted in by the proper control gains selected regardless of the presence of the disturbances. Perfectly selecting the gains for a good transient performance and maintaining high-precision control results for divergent working conditions in the real-time control is not a trivial work.

3.2 Auto gain-tuning rules

To effectively support gain selection for users, a simple strategy for gain tuning is employed: the control gains $\left(k_i|_{i\triangleq 1..5}\right)$ are separated into two terms: nominal elements $\left(\bar{k}_i|_{i\triangleq 1..5}\right)$ and variation elements $\left(\breve{k}_i\Big|_{i\triangleq 1..5}\right)$. The nominal ones play a key role in ensuring stability of the closed-loop system. The variation gains are self-adjusted to suppress unpredictable disturbances for the expected transient performance.

Furthermore, to ensure high control quality by avoiding sudden change of the gain variation, which could activate a chattering problem [25], the following constraints are noted.

Assumption 3: The variation terms $\left(\breve{k}_i\Big|_{i\triangleq 1..5}\right)$ and their first-order time derivatives are bounded.

Under operation of the flexible gains, the nonlinear control signal Eq. (8) is modified:

$$u_{new} = u + \frac{\text{sat}\left(\breve{k}_1\right)e_1}{a_2} \tag{11}$$

where $\text{sat}\left(\breve{k}_i\right)\Big|_{i\triangleq 1..5}$ are the saturation functions limited by upper-bound values $\left(\breve{k}_{i\_up}\Big|_{i\triangleq 1..5}\right)$ and lower-bound values $\left(\breve{k}_{i\_lo}\Big|_{i\triangleq 1..5}\right)$ as follows:

$$\text{sat}\left(\breve{k}_i\right)\Big|_{i\triangleq 1..5} = \begin{cases} \breve{k}_{i\_up} \text{ if } \left(\breve{k}_i \geq \breve{k}_{i\_up} \geq 0\right) \\ \breve{k}_i \quad \text{if } \left(\breve{k}_{i\_low} < \breve{k}_i < \breve{k}_{i\_up}\right) \\ \breve{k}_{i\_low} \text{ if } \left(\breve{k}_i \leq \breve{k}_{i\_low} \leq 0\right) \end{cases}$$

Lemma 2:

If a closed-loop system satisfies *Lemma 1*, it is stable for the time-varying gains complying with *Assumption 3*, and

$$\begin{cases} \Bigg|_{i=1..5} \begin{cases} 0 < k_{i\min} \leq k_i \leq k_{i\max} < \infty \\ \Delta_{\dot{\breve{k}}_i} < \infty \end{cases} \\ \left(\Delta_{k_1} + \Delta_{\breve{k}_1}\right) < k_{3\min}. \end{cases} \tag{12}$$

Proof of *Lemma 2* is given in *Appendix D*.

To comply with *Assumption 3*, the learning laws for the dynamic gains is structured from activation functions of the state control errors and leakage functions, which make sure boundedness of the learning gains.

The learning rules for the variation gains are proposed as follows:

$$\begin{cases} \dot{\breve{k}}_1 = -\sigma_1 e_1 \varepsilon - \text{sat}\left(\breve{k}_1\right) \\ \dot{\breve{k}}_2 = \dfrac{1}{\sigma_2}\varepsilon^2 - \eta_2 \text{sat}\left(\breve{k}_2\right) \\ \dot{\breve{k}}_3 = \dfrac{1}{\sigma_3} e_1 \varepsilon - \eta_3 \text{sat}\left(\breve{k}_3\right) \\ \dot{\breve{k}}_4 = \dfrac{1}{\sigma_4}\varepsilon\varphi - \eta_4 \text{sat}\left(\breve{k}_4\right) \\ \dot{\breve{k}}_5 = \dfrac{1}{\sigma_5}\text{ sgn }(e_1)\varphi - \eta_5 \text{sat}\left(\breve{k}_5\right) \end{cases} \tag{13}$$

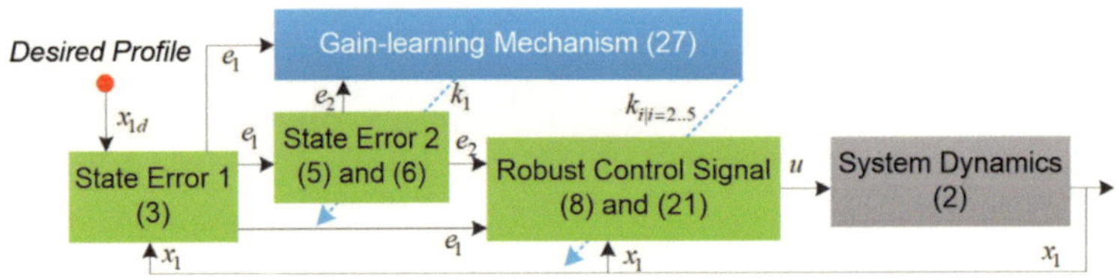

Figure 1.
Overview of the gain-learning backstepping controller for low-level subsystems.

where $\eta_{i|i=2..5}$ and $\sigma_{i|i=1..5}$ are positive learning rates.

To investigate the control performance of the learning control system, a new theorem is given.

Theorem 2:

If applying the control gains updated using Eq. (13) to a closed-loop system satisfying *Lemma 2*, asymptotic convergences of the state control error and variation gains are obtained.

Proof of *Theorem 2* could be referred in *Appendix E*.

Remark 4: Overview of the proposed controller is sketched in **Figure 1**. As stated in Theorem 1, the stability of the closed-loop system is ensured in a robust control framework, and as proven in Theorem 2, the adaptation of the control structure is highlighted by all the control gains learning for minimizing the tracking control error. The form of Eq. (E.4) reveals that the learning rates ($\sigma_{i|i=1..5}$ and $\eta_{i|i=2..5}$) can be employed with predefined values for specific control hardware.

Remark 5: In real-time applications, the proposed algorithm will be deployed in a discrete-time environment, the control errors will converge to arbitrary vicinities around zero. The desired control range can be however minimized under the learning mechanism proposed.

4. Real-time experiments

4.1 Setup

In this section, control performance of the intelligent controller is discussed based on verification results carried out in a real-time legged 2DOF robot. The experimental leg included one hip joint and one knee joint which were actuated by two BLDC motors. The mechanical design and a photograph of the actual leg are presented in **Figure 2**.

Incremental encoders were used to measure the joint angles, while a force sensor was placed in the shank of the robot to evaluate the ground contact force. The velocity signal was calculated from filtered backward differentiation of the position data. The robot was setup to freely move in both x and y directions. Total weight of the robot was about 15.74 kg. The proposed control algorithm was deployed in a NI Electrical Controller throughout LABVIEW software with a sampling time of 2 ms. The time derivative and integral terms in real-time implementation were approximated by Euler backward methods.

Two systematic parameters $\left(a_{1j}, a_{2j}\right)_{j=h,k}$ of the low-level systems could be estimated offline or online using a model-based identification method derived in previous works [27, 31, 32]. Nominal values of the parameters were approximately determined as $a_{1h} = 2.5; a_{2h} = 12.25; a_{1k} = 0.5; a_{2k} = 15$.

4.2 Comparative control results

Both the hip and knee joints were controlled at the same time using the same control algorithm proposed. The controller was also compared with an adaptive robust extended-state-observer-based (ARCESO) controller, a robust integral-sign-error (RISE) controller, and another case of itself with fixed gains (nominal gains) in Eq. (8), which is denoted as the robust backstepping (RB) controller.

The ARCESO controller was designed based on a previous work [30] wherein their control gains were chosen as

$$\begin{cases} k_{1h} = 100; k_{2h} = 100; \omega_{0h} = 60; \theta_{\min h} = [6, 1]^T; \theta_{\max h} = [25, 5]^T; \hat{\theta}_h(0) = [12.25, 2.5]^T; \\ k_{1k} = 80; k_{2k} = 75; \omega_{0k} = 40; \theta_{\min k} = [5, 0.2]^T; \theta_{\max k} = [30, 1.5]^T; \hat{\theta}_k(0) = [15, 0.5]^T; \end{cases}$$

The RISE controller was implemented based on a robust integral theory [13] to control the studied system Eq. (2) without considering the measurement noise (υ). Its control signal was:

$$\begin{cases} e_1 = x_1 - x_{1d}; e_2 = \dot{e}_1 + k_{RISE1} e_1 \\ u_{RISE} = -a_2^{-1} \left(-a_1 x_2 - \ddot{x}_{1d} + k_{RISE1}(x_2 - \dot{x}_{1d}) + k_{RISE2} e_2 + \int_0^t (k_{RISE3} e_2 + k_{RISE4} \operatorname{sgn}(e_2)) d\tau \right) \end{cases}. \tag{14}$$

The RISE control gains were set to be:

$$\begin{cases} k_{RISE1h} = 53; k_{RISE2h} = 85.4; k_{RISE3h} = 20; k_{RISE4h} = 250; \\ k_{RISE1k} = 45; k_{RISE2k} = 65; k_{RISE3k} = 17; k_{RISE4k} = 235; \end{cases}$$

The nominal gains of the proposed controllers were chosen to be:

$$\begin{cases} \bar{k}_{1h} = 1.5; \bar{k}_{2h} = 72.5; \bar{k}_{3h} = 2000; \bar{k}_{4h} = 50; \bar{k}_{5h} = 200; \\ \bar{k}_{1k} = 5; \bar{k}_{2k} = 72.5; \bar{k}_{3k} = 2000; \bar{k}_{4k} = 50; \bar{k}_{5k} = 200; \end{cases}$$

The excitation signals (c and φ) of the learning laws Eq. (13) were directly synthesized from the control error (e_1) and its high-order time derivatives based on Eq. (D.1):

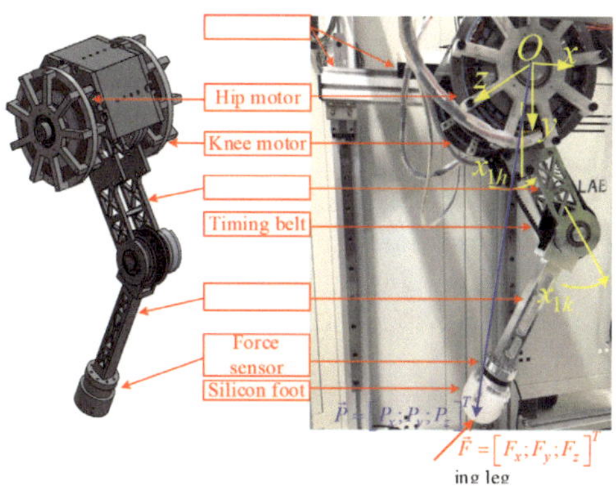

Figure 2.
Design and setup of the experimental testing system.

$$\begin{cases} \varepsilon = \dot{e}_1 + k_1 e_1 \\ \varphi = (\ddot{e}_1 + k_2 \dot{e}_1) + \left(k_3 + (k_2 - k_1)k_1 - \text{sat}\left(\breve{k}_1\right)\right)e_1 \end{cases} \tag{15}$$

From the nominal control gains selected, the feasible ranges of the variation gains were then chosen to gratify the constraint Eq. (9):

$$\begin{cases} \begin{cases} \breve{k}_{1h\,\min} = -1.4; \breve{k}_{2h\,\min} = -66.5; \breve{k}_{3h\,\min} = -1000; \breve{k}_{4h\,\min} = -49; \\ \breve{k}_{1h\,\max} = 3.5; \breve{k}_{2h\,\max} = 500; \breve{k}_{3h\,\max} = 4000; \breve{k}_{4h\,\max} = 1500; \breve{k}_{5h\,\min} = -199; \breve{k}_{5h\,\max} = 1500; \end{cases} \\ \begin{cases} \breve{k}_{1k\,\min} = -4.9; \breve{k}_{2k\,\min} = -66.5; \breve{k}_{3k\,\min} = -1000; \breve{k}_{4k\,\min} = -49; \\ \breve{k}_{1k\,\max} = 5; \breve{k}_{2k\,\max} = 500; \breve{k}_{3k\,\max} = 4000; \breve{k}_{4k\,\max} = 1500; \breve{k}_{5k\,\max} = 1500; \breve{k}_{5k\,\min} = -199; \end{cases} \end{cases}$$

The learning rates ($\sigma_{i|i=1..5}$ and $\eta_{i|i=2..5}$) were then set to comply with the condition Eq. (12) and to ensure the variation gains freely varying inside their predetermined ranges. For simplicity, the relaxation rates ($\eta_{i|i=2..5}$) could be chosen to be *1* or *2*. Finally, the rates tuned were as

$$\begin{cases} \sigma_{1h} = 1000; \sigma_{2h} = 10^{-2}; \eta_{2h} = 1; \sigma_{3h} = 2 \times 10^{-4}; \eta_{2h} = 1; \sigma_{4h} = 0.05; \\ \eta_{4h} = 2; \sigma_{5h} = 0.03; \eta_{5h} = 1; \\ \sigma_{1k} = 1000; \sigma_{2k} = 10^{-1}; \eta_{2k} = 1; \sigma_{3k} = 2 \times 10^{-3}; \eta_{2k} = 1; \sigma_{4k} = 0.05; \\ \eta_{4k} = 2; \sigma_{5k} = 0.03; \eta_{5k} = 1; \end{cases}$$

4.2.1 Simple verification

In this validation series, the proposed controller was only applied for position-tracking control of the hip joint. A sinusoidal signal of $x_{1dh} = 14\sin(4\pi t)$ (deg) was chosen as the desired trajectory of the test. The leg was put to move freely in the air to eliminate the external disturbance. **Figure 3(a)** presents the experimental data obtained by the comparative controllers. The ARCESO controller produced a very small control error of ±0.14 deg. (~1.0%) in the high-speed tracking control thanks to the use of an effective adaptive-disturbance learning mechanism. The ARCESO control performance was still however limited with fast-variation disturbances [30]. By adopting the integral-robust control signal Eq. (14) to compensate for the lumped disturbance (d) in the low-level system Eq. (2), the RISE controller also exhibited a high control accuracy (control error: [−0.16; 0.14] deg. (~1.14%)). In fact, in real-time applications, improper control gains selected or large measurement noise (v) could degrade the RISE control performance. As operating under the highly robust design Eq. (8) against all the disturbances, the RB technique provided better control precision (control error: ±0.138 deg. (~ 0.98%)). Theoretically, the control performance could be further increased if the best control gains were found, but it may be a time-consuming work. As a solution, the gain-tuning process could be supported by the learning mechanism Eqs. (11) and (13) proposed. Indeed, the control quality was intuitively enhanced by applying GARB control method, which yielded the smallest control error of ±0.085 deg. (~0.6%).

The gain-learning behaviors are illustrated in **Figure 3(b)**. As seen in the figure, the variation gains were automatically changed in various ways under the adaptation laws to minimize the control error. The maximum-absolute (MA) and root-mean-squares (RMS) values of the control errors from after system was stable (from 2 s to 5 s) are summarized in **Table 1**. Herein, the proposed controller shows outperformance as comparing to the previous methods.

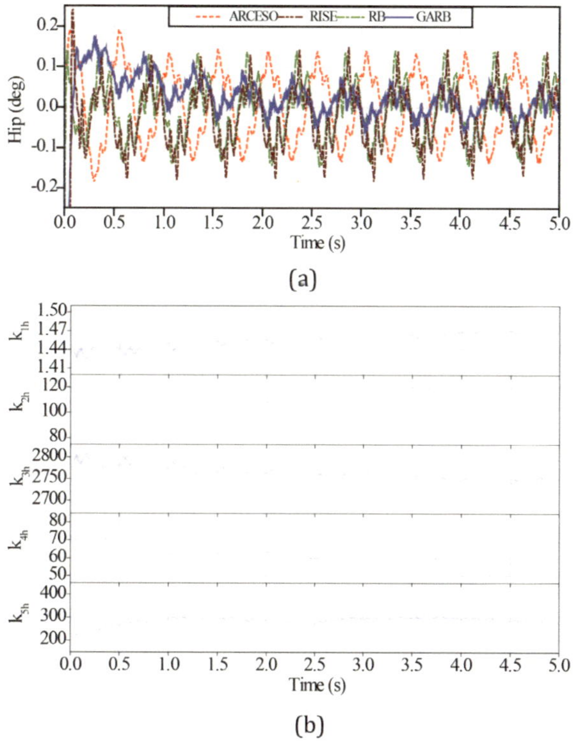

Figure 3.
Experimental results of the single-joint test. (a). Comparative control errors of the testing controllers. (b). Gain learning of the GARB controller.

Control error	ARCESO	RISE	RB	GARB
MA	0.140	0.160	0.138	0.085
RMS	0.080	0.074	0.072	0.030

Table 1.
Performance comparison of the controllers for the single-joint validation.

4.2.2 Complex verification

To deeper challenge to the special properties of the proposed controller, the robot was controlled to perform a squatting exercise in three different working cases: in the air, on the ground, and with ground contact. The frequency and amplitude of the squatting motion were selected to be 2 Hz and 80 mm, respectively. These tests are normal working cases of the leg in real-time missions. The desired trajectories (x_{1dh} and x_{1dk}) of the two robot joints (hip and knee) are plotted in **Figure 4**. The trajectories were derived from desired foot motion of

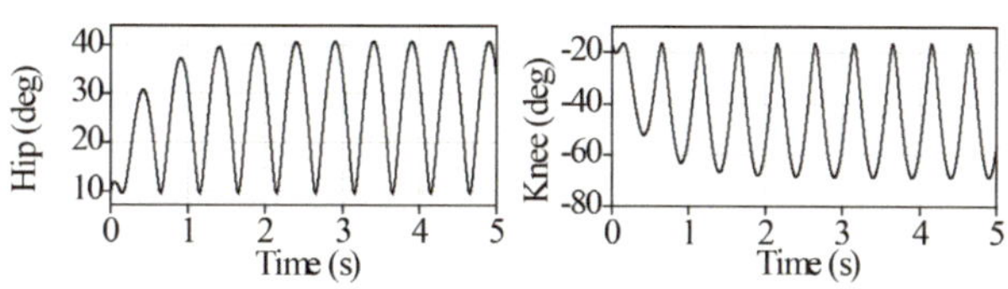

Figure 4.
Desired profiles of the robot joints in the multiple-joint tests.

$(P_y = 460 + 40\sin(4\pi t); P_x = 0)$mm using simple inverse-kinematics computation as noted in *Appendix F*.

4.2.2.1 Verification with minor external disturbances

Although the robot worked in the air, the disturbances affecting the control joints were large due to high-speed control and interaction forces between the joints during the system movement. The dynamical and statical control results obtained by the validated controllers are respectively shown in **Figure 5** and **Table 2**. In spite of operating with faster motions (192.2 (deg/s) and 324.8 (deg/s) for the hip and knee joints) and in harder internal disturbance conditions, the ARCESO controllers

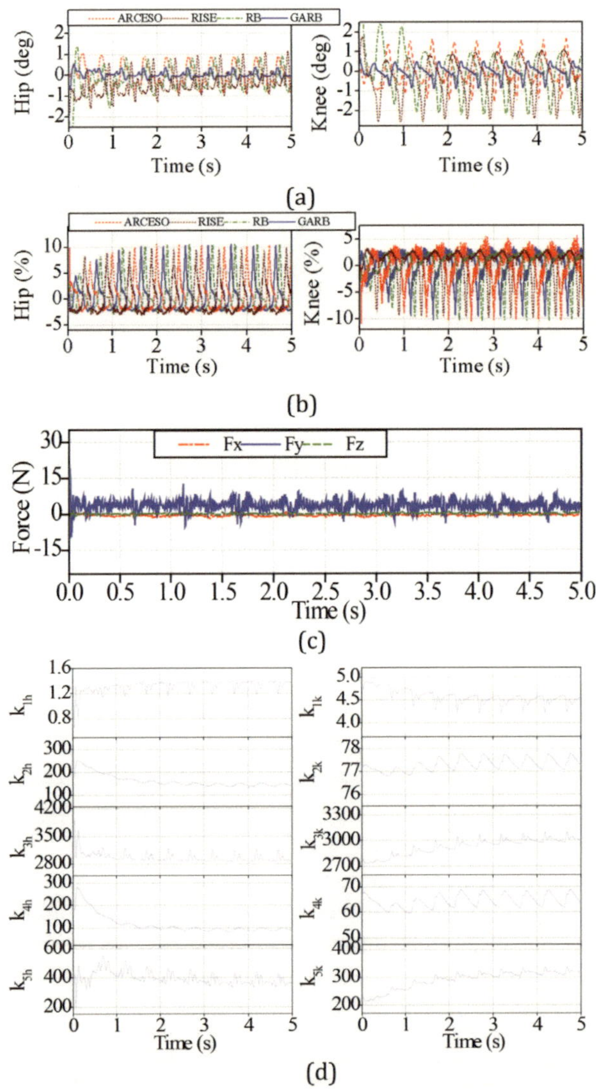

Figure 5.
Experimental results of the testing controllers for the multiple-joint test in case of small external disturbance. (a). Control errors of the comparative controllers. (b). Control inputs generated by the comparative controllers. (c) Forces measured at the shank with respect to the GARB controllers. (d) Gain learning of the GARB controllers.

	Control error	ARCESO	RISE	RB	GARB
HIP	MA	0.8	1.1	1	0.35
	RMS	0.5	0.78	0.75	0.17
KNEE	MA	1.5	2.3	2	0.8
	RMS	0.74	1.17	1.15	0.37

Table 2.
Performance comparison of the validated controllers in the small disturbance tests.

maintained high control outcome thanks to the strong adaptation ability: $\pm 0.8\text{deg}(\sim 4.8\%)$ and $\pm 1.5\text{deg}(\sim 4.1\%)$for the hip and knee joints.

As seen in **Figure 5(a)**, the robust backstepping designs coped with the reaction forces as well. The RB and RISE controllers stabilized the control errors inside acceptable ranges: the errors for hip joint and knee joint are respectively $[-0.5 \rightarrow 1]\text{deg}(\sim 6\%)$and $[-2 \rightarrow 1]\text{deg}(\sim 5.5\%)$ with the RB, while those are $[-0.8 \rightarrow 1.1]\text{deg}(\sim 6.6\%)$ and $[-2.3 \rightarrow 1.2]\text{deg}(\sim 6.3\%)$ with the RISE. In the new working conditions, excellent control errors were also resulted in by the GARB controller based on a new set of the control gains found. **Figure 5(d)** depicts the variation gains that were incorporated with the proposed robust design Eq. (11) to create a better control performance as comparing to the others ($[-0.2 \rightarrow 0.35]\text{deg}(\sim 2.1\%)$and $[-0.8 \rightarrow 0.5]\text{deg}(\sim 2.2\%)$ for the hip and knee control errors).

Comparison of the control power required to conduct the high-speed control motions is shown in **Figure 5(b)**. Although the control efforts of the controllers were almost same for this mission. Only minor disparate nonlinearities in the control signals would lead to the divergence on control performances. The figure also reveals that the GARB controllers generated applicable control inputs even though the learning gains were moderated in a risk of the high-order measurement noise. The benefit comes from the low-pass-filter like nature of the gain-learning algorithm proposed. External force affecting the leg measured in the shank using the GARB controllers is presented in **Figure 5(c)**. The coordinate of the measured force is sketched in **Figure 2**. This experiment shows the higher control accuracies and demonstrates the advantages of the proposed controller as comparing to other controllers.

4.2.2.2 Verification with large external disturbances

In this experiment, the robust adaptive ability of the proposed controller was harshly investigated under conditions of heavy external load. The robot was put on the ground and supported by sliders in both the x and y directions. To avoid damage for the robot, only the proposed controller was used in the verification. The control results obtained are plotted in **Figure 6**. In this test, the external forces reacting from environment were significantly increased from 10 N to 390 N. The data presented in **Figure 6(a)** however implies that the controller still provided acceptable control accuracy: $[-0.35 \rightarrow 0.68]\text{deg}(\sim 4.08\%)$ and $[-1.5 \rightarrow 1.1]\text{deg}(\sim 4.1\%)$ for the hip and knee joints.

As demonstrated in **Figure 6(b)**, in this case the system used larger energy than in the second one to execute the fast-tracking control under critical conditions. As presented in **Figure 6(d)**, the control gains were also automatically changed to higher values to deal with large disturbances for a smallest possible control error. Hence, the strong robustness and fast adaptability of the proposed method can be confirmed via this investigation.

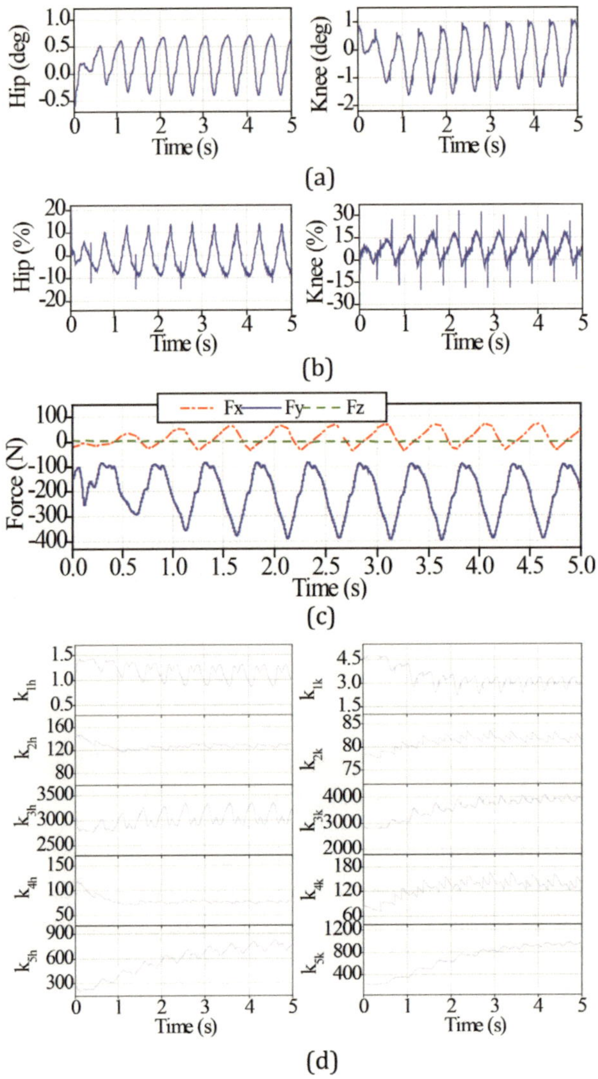

Figure 6.
Experimental results of the GARB controllers for the multiple-joint test in case of large external disturbance. (a) Control errors of the GARB controllers. (b) Control inputs generated by the GARB controllers. (c) Measurement of ground reaction forces. (d) Gain learning of the proposed mechanisms.

4.2.2.3 Verification with fast-variation external disturbances

In this case study, transient behaviors of the designed controller were carefully validated by using fast-variation external disturbances. The robot was still controlled to conduct the same squatting work. Harder testing conditions were constituted by two consecutive distinguished phases of one working cycle: a ground-contact phase and ground-release phase. **Figure 7(c)** shows the ground-reaction forces measured during the test. The nature of the external disturbance in this case was different from those in the previous cases. Fast variation of the reaction forces may make the system instable. The control system designed had however showed the concrete robustness and impressive adaptation in real-time control again.

As presented in **Figure 7(a)**, the closed-loop system provided good performance: $[-0.22 \rightarrow 0.76]\,\text{deg}\,(\sim 4.56\%)$ for the hip joint and $[-0.8 \rightarrow 0.7]\,\text{deg}\,(\sim 2.2\%)$ for the knee joint. The control energy and control

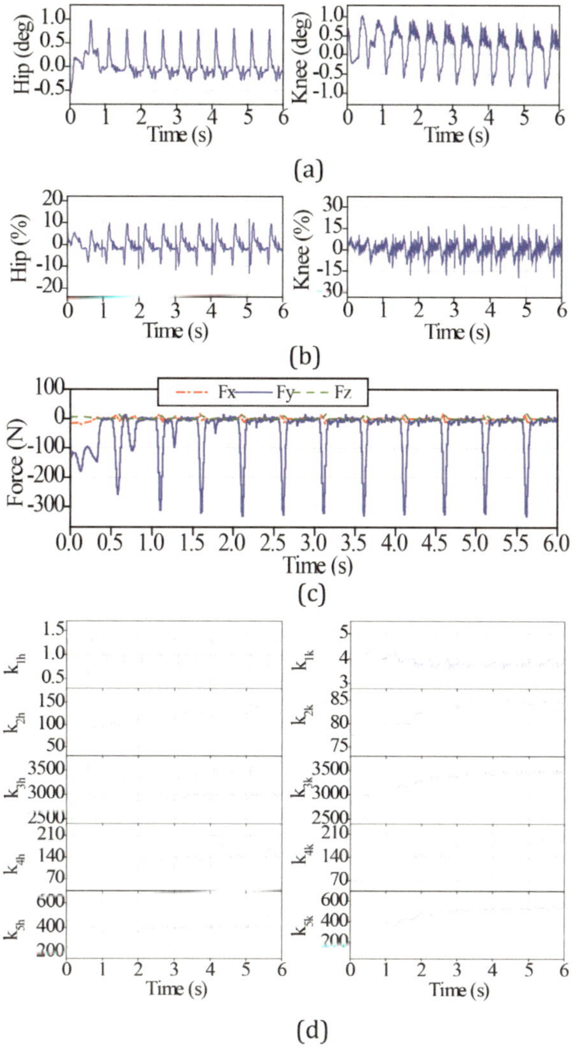

Figure 7.
Experimental results of the GARB controllers for the multiple-joint test in case of fast-variation external disturbance. (a) Control errors of the GARB controllers. (b) Control inputs generated by the GARB controllers. (c) Measurement of ground reaction forces. (d) Gain learning of the proposed mechanisms.

parameters were varied to properly adapt to change of the new working conditions. **Figure 7(d)** shows that the new ranges of the control gains were found by the proposed algorithm, and **Figure 7(b)** presents the required energy for the new test.

4.2.3 Additional Statical note

The RMS values of the control errors, control signals (u), and the ground-reaction forces for the hip and knee joints of the complex validation process are noted in **Table 3**. The data imply that the GARB controller was able to result in good control performances with the preset learning rates in the high-speed task under different working conditions. The learning mechanism and robust control technique generated proper power for each test case to effectively realize the control objective. Some snapshots of the robot movement in the last experiment are shown in **Figure 8**.

		HIP		KNEE	
Testing condition	**Force (N)**	**Error (deg)**	***u* (%)**	**Error (deg)**	***u* (%)**
Small external disturbance	4.1	0.168	3.149	0.327	2.932
Large external disturbance	229.8	0.405	6.4653	0.782	9.7454
Fast-variation external disturbance	89.4	0.267	3.888	0.496	4.491

Table 3.
Performance comparison of the garb controllers in the multiple-joint tests.

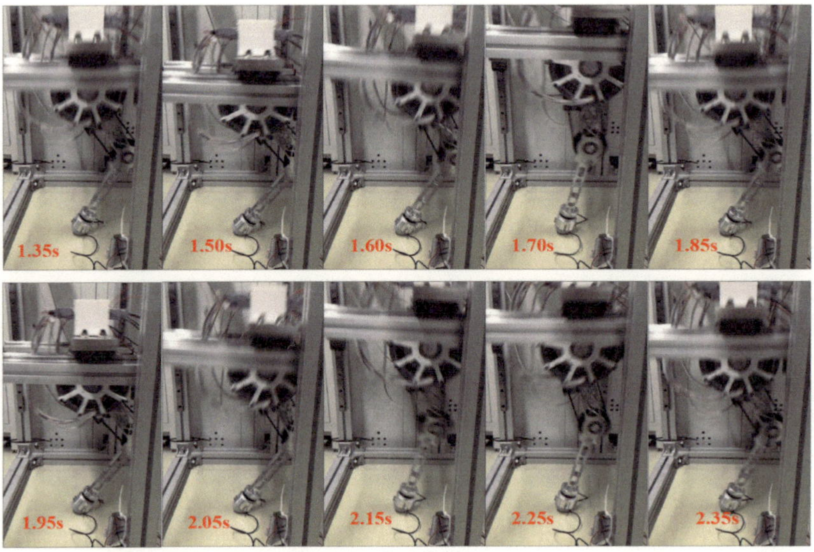

Figure 8.
Snapshots of the leg motion in the large external disturbance test.

5. Discussion

In many humanoid robots, so far, one mainly focuses on building complicated high-level control structures while in the low-level framework simple controllers, such as PID or SMC, were normally employed to realize the given command [28, 33]. Obviously, to ensure the whole system operate as expected, auto-adjusting terms must be implemented at the upper-level framework to compensate for the imperfection of the simple low-level actions [34, 35]. With such the cross-over interference between the control layers, it was hard to provide high accuracies and fast responses for the overall system [28, 36]. Indeed, in our real-time experiments with the legged robot, well-tuned PID controllers could be adopted for squatting tests in a certain case. When the working condition changed, the control system could be damaged by the PID controller due to degradation of the control performance. Of course, precision controllers could be employed in the low-level layer but their simplicity in implementation and less computation burden should be preserved. The gain-adaptive robust backstepping control algorithm has been developed in comply with these strict requirements.

As noted in the control signal Eq. (8), if one chooses $k_5 = 0$ and $a_1 = 0$, the nonlinear control method becomes an ordinary PID controller. In another sense, if the control gains k_4 and k_5 are removed, the control signal Eq. (8) presents for a conventional form of the SMC scheme in which e_2 is the sliding surface. Hence, users have various options in adoption of the designed controller, which could be easily switched to basic control options [6, 8, 28, 35].

Note also that the input gain constant (a_2) could be selected with an arbitrarily positive constant while the nominal dynamical constant (a_1) could be zero or any bounded value. Their deviations could be counted into the lumped disturbance (d) or extended disturbance (h). One possible way to determine such the terms is use of the model-based identification method presented in previous works [27, 30, 31].

As comparing to other intelligent gain-learning algorithms such as neural network or fuzzy logic engines, the computational burden and fast response are noteworthy advantages [9–11, 37, 38]. However, in some cases, one does not need to use the nominal dynamics or ($a_1 = 0$), and at that time, overall design of the proposed control method becomes a model-free controller.

The experimental results have confirmed the outperformance of the gain-learning controllers over other robust adaptive nonlinear controllers, such as ARCESO and RISE [13, 30], thanks to a high-degree-of-learning mechanism. Furthermore, the designed controller has been improved from the former controller [27] to increase the real-time applicability by removing third-order time-derivation terms in the control signal.

From the above analyses, the flexibility of the designed controller in terms of working efficiency and user implementation are intuitively observed. Its feasibility in movable robots have been also confirmed by intensive experiments.

6. Summary

This chapter presents a gain-adaptive robust position-tracking controller for low-level subsystems of large robotic systems. The mathematical model of the system dynamics was reviewed to provide necessary information for the controller design. To realize the tracking control objective, a robust control signal based on the backstepping scheme was adopted. In fact, this design is a nonlinear extension of ordinary PID controller or conventional sliding mode controller. New adaptation laws were developed to automatically tune the control gains for different working conditions. The learning mechanism was activated by various forms of the control error and deactivated by the relaxation functions.

Stability of the overall system was concretely maintained by proper Lyapunov-based constraints. Extended real-time experiments were conducted to verify the performance of the proposed controller. The results achieved confirmed the advantages on the robustness, adaptation, high accuracy, and fast response of the proposed controller. Depending on the usage purpose of user, the controller could be simplified to become a gain-learning PID controller or an adaptive robust sliding mode controller.

Appendix A. Proof of Lemma 1

Let define the following new disturbance:

$$h = -\ddot{x}_{1d} + \dot{\upsilon} + k_2\upsilon + d \tag{A.1}$$

Also synthesize a new state variable and lumped term as follows:

$$\begin{cases} \varphi = -\int_0^t ((k_4\varepsilon + k_5 \operatorname{sgn}(e_1))d\varsigma) + h \\ \varepsilon = \upsilon + e_2 \end{cases} \tag{A.2}$$

By noting Eqs. (3), (4), (8), and (A.2), the following dynamics are obtained:

$$\begin{cases} \dot{e}_1 = \varepsilon - k_1 e_1 \\ \dot{\varepsilon} = -k_3 e_1 - (k_2 - k_1)\varepsilon + \varphi \end{cases} \tag{A.3}$$

The following positive function is studied:

$$V_{10} = 0.5k_4k_3e_1^2 + 0.5k_4\varepsilon^2 + 0.5\varphi^2 + \int_0^t \left(k_5 \operatorname{sgn}(e_1) - \dot{h}\right)\dot{\varepsilon}d\tau + V_{100} \tag{A.4}$$

where V_{100} is a positive constant selected as.

$$V_{100} = \frac{0.5\left(k_5 + \Delta_{\dot{h}}\right)^2}{k_4} + \left(k_5 + \Delta_{\dot{h}}\right)|\varepsilon(0)| \tag{A.5}$$

Here, $\Delta_{\bullet} = \max(|\bullet|)$ is the maximum absolute value of function $(\bullet)$.

The proof of the positive function V_{10} can be obtained by applying integral inequalities and the condition Eq. (A.5).

The time derivative $\dot{V}_{10}$ is simplified using combinations of Eqs. (A.2) and (A.3), as follows:

$$\begin{aligned} \dot{V}_{10} &= k_4k_3e_1(\varepsilon - k_1e_1) + k_4\varepsilon\dot{\varepsilon} + \left(k_5 \operatorname{sgn}(e_1) - \dot{h}\right)\dot{\varepsilon} \\ &-(\dot{\varepsilon} + (k_2 - k_1)\varepsilon + k_3e_1)\left(k_4\varepsilon + k_5 \operatorname{sgn}(e_1) - \dot{h}\right) \\ &\le -k_3k_5|e_1| - k_3k_4k_1\left(|e_1| - \frac{\Delta_{\dot{h}}}{2k_4k_1}\right)^2 + \frac{k_3\Delta_{\dot{h}}^2}{4k_4k_1} \\ &-(k_2 - k_1)k_4\left(|\varepsilon| - \frac{\left(k_5 + \Delta_{\dot{h}}\right)}{2k_4}\right)^2 + \frac{(k_2 - k_1)\left(k_5 + \Delta_{\dot{h}}\right)^2}{4k_4} \end{aligned} \tag{A.6}$$

Let define the following positive constants:

$$\begin{cases} e_{\dagger} = \dfrac{\Delta_{\dot{h}}}{2k_4k_1} + \sqrt{\dfrac{1}{k_3k_4k_1}\left(\dfrac{k_3\Delta_{\dot{h}}^2}{4k_4k_1} + \dfrac{(k_2 - k_1)\left(k_5 + \Delta_{\dot{h}}\right)^2}{4k_4}\right)} \\ \varepsilon_{\dagger} = \dfrac{\left(k_5 + \Delta_{\dot{h}}\right)}{2k_4} + \sqrt{\dfrac{1}{(k_2 - k_1)k_4}\left(\dfrac{k_3\Delta_{\dot{h}}^2}{4k_4k_1} + \dfrac{(k_2 - k_1)\left(k_5 + \Delta_{\dot{h}}\right)^2}{4k_4}\right)} \end{cases}$$

By noting *Assumption 2*, the terms $\Delta_{\dot{h}}, e_{\dagger}, \varepsilon_{\dagger}$ are bounded. If $(|e_1| > e_{\dagger})$ and/or $(|\varepsilon| > \varepsilon_{\dagger})$, $\dot{V}_{10}$ is negative. It implies e_1 and ε are bounded [19, 29]. Therefore, *Lemma 1* is proven. ■

B. Proof of Theorem 1

A new Lyapunov function is investigated.

$$V_{11} = V_{10} + P_1(t) \tag{B.1}$$

where $P_1(t)$ is a positive function defined as:

$$\begin{cases} P_1(t) = P_1(0) + \int_0^t \left(\left(k_5 \operatorname{sgn}(e_1) - \dot{h}\right)((k_2 - k_1)\varepsilon + k_3 e_1)\right) d\tau \\ P_1(0) = 2(k_2 - k_1)\Delta_e\left(k_5 + \Delta_{\dot{h}}\right) \end{cases} \tag{B.2}$$

The proof of the function $P_1(t)$ can be referred in *Appendix C*.
The time derivative of the Lyapunov function in adoption of Eqs. (A.6) and (B.2) is.

$$\begin{aligned} \dot{V}_{11} &= -k_3 k_4 k_1 e_1^2 - k_4(k_2 - k_1)\varepsilon^2 + \dot{P}_1(t) - \left(k_5 \operatorname{sgn}(e_1) - \dot{h}\right)((k_2 - k_1)\varepsilon + k_3 e_1) \\ &= -k_3 k_4 k_1 e_1^2 - k_4(k_2 - k_1)\varepsilon^2 \end{aligned} \tag{B.3}$$

From Eqs. (2), (9)–(10), (A.6), (B.1)–(B.3), and *Assumptions 1* and *2*, we have: $[e_1, \varepsilon]^T \in L_2^2$. By recalling (A.4), $\dot{e}_1$ is bounded, and.

$$\ddot{\varepsilon} = -(k_2 - k_1)\dot{\varepsilon} + k_1 k_3 e_1 + \dot{h} - (k_3 + k_4)\varepsilon - k_5 \operatorname{sgn}(e_1) \tag{B.4}$$

It implies that $\dot{\varepsilon}$ is bounded. Hence, by using Barbalat's lemma [39], *Theorem 1* is proven. ■

C. Proof of the positive function P1(t)

The function $P_1(t)$ expressed in Eq. (B.2) can be expanded using the error dynamics Eq. (A.3) and integral inequalities as follows:

$$\begin{aligned} P_1(t) \geq P_1(0) &+ \int_0^t \left(\left(k_5 \operatorname{sgn}(e_1) - \dot{h}\right)\left(k_3 + k_1 k_2 - k_1^2\right)e_1\right) d\tau \\ &-(k_2 - k_1)\int_0^t \left(\dot{h}\dot{e}_1\right) d\tau \quad k_5(k_2 - k_1)\int_{e_1(0)}^{e_1(t)} d(e_1) \end{aligned} \tag{C.1}$$

By applying the integrating procedures in previous works [8] and comparison inequality, we have.

$$P_1(t) \geq P_1(0) + \int_0^t \left(\left(k_5 - \Delta_{\dot{h}}\right)\left(k_3 + k_1 k_2 - k_1^2\right)|e_1|\right) d\tau \ - (k_2 - k_1)\left(\Delta_{\dot{h}} + k_5\right)|e - e(0)| \tag{C.2}$$

The proof is completed by noting *Lemma 1*, the conditions Eqs. (10) and (B.2), the definition Eq. (A.1), and *Assumptions 1* and *2*.

D. Proof of Lemma 2

By applying the control input Eq. (11) to the dynamics Eq. (2), the closed-loop system is:

$$\begin{cases} \dot{e}_1 = \varepsilon - k_1 e_1 \\ \dot{\varepsilon} = -(k_2 - k_1)\varepsilon - \left(k_3 - \text{sat}\left(\breve{k}_1\right) - \dot{\breve{k}}_1\right)e_1 + \varphi. \end{cases} \tag{D.1}$$

A new positive function is studied.

$$V_{20} = 0.5\varphi^2 + \int_0^t \left(k_5 \,\text{sgn}\,(e_1) - \dot{h}\right)\dot{\varepsilon}d\tau + \int_0^t k_4\left(k_3 - \text{sat}\left(\breve{k}_1\right) - \dot{\breve{k}}_1\right)e_1\dot{e}_1 d\tau + \int_0^t k_4\varepsilon\dot{\varepsilon}d\tau + V_{200} \tag{D.2}$$

where V_{200} is a positive constant selected as.

$$V_{200} = \frac{0.5\left(k_{5\max} + \Delta_{\dot{h}}\right)^2}{k_{4\min}} + \left(k_{5\max} + \Delta_{\dot{h}}\right)|\varepsilon(0)| + 0.5k_{4\max}(\varepsilon(0))^2 + 0.5k_{4\max}\left(k_{3\max} + \Delta_{\breve{k}_1} + \Delta_{\dot{\breve{k}}_1}\right)(e_1(0))^2 \tag{D.3}$$

The proof of *Lemma 1* can be reused for the positive function V_{20} based on Eq. (D.3) and for its time derivative. Then, the time derivative of the new function is:

$$\begin{aligned} \dot{V}_{20} &= \left(k_5 \,\text{sgn}\,(e_1) - \dot{h}\right)\dot{\varepsilon} + k_4\left(k_3 - \text{sat}\left(\breve{k}_1\right) - \dot{\breve{k}}_1\right)e_1(\varepsilon - k_1 e_1) + k_4\varepsilon\dot{\varepsilon} \\ &\quad -\left(\dot{\varepsilon} + (k_2 - k_1)\varepsilon + \left(k_3 - \text{sat}\left(\breve{k}_1\right) - \dot{\breve{k}}_1\right)e_1\right)\left(k_4\varepsilon + k_5 \,\text{sgn}\,(e_1) - \dot{h}\right) \\ &\leq -|e_1|\left(k_3 - \text{sat}\left(\breve{k}_1\right) - \dot{\breve{k}}_1\right)\left(k_4 k_1|e_1| + k_5 - \Delta_{\dot{h}}\right) - (k_2 - k_1)|\varepsilon|\left(k_4|\varepsilon| - k_{5\max} - \Delta_{\dot{h}}\right) \end{aligned} \tag{D.4}$$

By employing the same discussion with *Lemma 1* under *Assumption 2, Lemma 2* is proven.■

E. Proof of Theorem 2

Let consider the following Lyapunov function:

$$V_2 = 0.5\left(\bar{k}_3\bar{k}_4 e_1^2 + \bar{k}_4\varepsilon^2 + 2P_2(t) + \varphi^2 + \sigma_2\bar{k}_4\breve{k}_2^2 + \sigma_3\bar{k}_4\breve{k}_3^2 + \sigma_4\breve{k}_4^2 + \sigma_5\breve{k}_5^2\right) \tag{E.1}$$

where $P_2(t)$ is a positive function which is chosen as follows:

$$P_2(t) = P_2(0) + \int_0^t \left(\left(\bar{k}_5 \,\text{sgn}\,(e_1) - \dot{h}\right)\varphi\right)d\tau_2 \tag{E.2}$$

The proof of the function $P_2(t)$ can be satisfactory using the similar arguments presented in *Appendix C* with the following conditions:

$$\begin{cases} P_2(0) = 2\left(\overline{k}_5 + \Delta_{\dot{h}}\right)\left(\Delta_{\dot{e}} + k_{2\max}\Delta_e\right) \\ \left(k_3 + (k_2 - k_1)k_1 - \text{sat}\left(\breve{k}_1\right)\right)_{\min} > 0 \\ \overline{k}_5 \geq \Delta_{\dot{h}} \end{cases} \tag{E.3}$$

Substituting Eqs. (D.1) and (13) to the time derivative of the new Lyapunov function leads to.

$$\begin{aligned} \dot{V}_2 = &-\overline{k}_3\overline{k}_4k_1e_1^2 - \left(\overline{k}_2 - k_1\right)\overline{k}_4\varepsilon^2 - \overline{k}_4\sigma_1(e_1\varepsilon)^2 - \eta_2\sigma_2\overline{k}_4\breve{k}_2\text{sat}\left(\breve{k}_2\right) \\ &-\eta_3\sigma_3\overline{k}_4\breve{k}_3\text{sat}\left(\breve{k}_3\right) - \eta_4\sigma_4\breve{k}_4\text{sat}\left(\breve{k}_4\right) - \eta_5\sigma_5\breve{k}_5\text{sat}\left(\breve{k}_5\right) \end{aligned} \tag{E.4}$$

Theorem 2 is proven by noting Eqs. (2), (D.1), (E.1), (E.4), *Assumptions 1* and *2*, and the discussions in the proof of *Theorem 1*. ■

F. Inverse Kinematics of the robot leg

The desired angles of the leg joints (hip (x_{1dh}) and knee (x_{1dk})) can be calculated from the position of the foot (the end-effector) using the following inverse kinematics:

$$\begin{cases} x_{1dh} = \text{atan2}\left(P_x, P_y\right) + \text{arc}\left(\dfrac{P_x^2 + P_y^2 + l_1^2 - l_2^2}{2l_1\sqrt{P_x^2 + P_y^2}}\right) \\ x_{1dk} = \text{atan2}(P_x - l_1\sin\left(x_{1dh}\right), P_x - l_1\cos\left(x_{1dh}\right)) - x_{1dh} \end{cases} \tag{E.5}$$

where $l_1 = 0.21\,\text{m}$ and $l_2 = 0.295\,\text{m}$ are the link lengths of robot (thigh and shank), respectively. P_x and P_y are the end effector position of the robot foot with respect to the robot coordinate setting at the hip joint, as sketched in **Figure 2(b)**. The feasible working range of the hip joint was selected to be $[0 \rightarrow +80]$ deg.

Author details

Dang Xuan Ba[1*] and Joonbum Bae[2]

1 HCMC University of Technology and Education (HCMUTE), Ho Chi Minh City, Vietnam

2 Ulsan National Institute of Science and Technology (UNIST), Ulsan City, South Korea

*Address all correspondence to: badx@hcmute.edu.vn

References

[1] S. Seok, A. Wang, M. Y. Chuah, D. J. Hyun, J. Lee, D. M. Otten, F. H. Lang, and S. Kim, "Design principles for energy-efficient legged locomotion and implementation on the MIT Cheetah robot," IEEE/ASME Trans. Mechatronics, vol. 20, no. 3, pp. 1117–1129, 2015.

[2] D. Das, N. Kumaresan, V. Nayanar, K. Navin Sam, and N. Ammasai Gounden, "Development of BLDC Motor-Based Elevator System Suitable for DC Microgrid," IEEE/ASME Trans. Mechatronics, vol. 21, no. 3, pp. 1552–1560, 2016.

[3] W. S. Huang, C. W. Liu, P. L. Hsu, and S. S. Yeh, "Precision control and compensation of servomotors and machine tools via the disturbance observer," IEEE Trans. Ind. Electron., vol. 57, no. 1, pp. 420–429, 2010.

[4] L. Lu, B. Yao, Q. F. Wang, and Z. Chen, "Adaptive robust control of linear motors with dynamic friction compensation using modified Lugre model," Automatica, vol. 45, no. 12, pp. 2890–2896, 2009.

[5] Y. Yang, Y. Wang, and P. Jia, "Adaptive robust control with extended disturbance observer for motion control of DC motors," Electronic Letters, vol. 51, no. 22, pp. 1761–1763, 2015.

[6] P. Roco, "Stability of PID control for industrial robot arms," IEEE Trans. Robot. Automation, vol. 12, no. 4, pp. 606–614, 1996.

[7] S. Skoczowski, S. Domesk, K. Pietrusewicz, and B. Broel-Plater, "A method for improving the robustness of PID control," IEEE Trans. Ind. Electron., vol. 58, no. 6, pp. 1669–1676, 2005.

[8] V. Mummadi, "Design of robust digital PID controller for H-bridge soft-switching boost converter," IEEE Trans. Ind. Electron., vol. 58, no. 7, pp. 2883–2897, 2011.

[9] R. J. Wai, J. D. Lee, and K. L. Chuang, "Real-time PID control strategy for maglev transportation system via particle swarm optimization," IEEE Trans. Ind. Electron., vol. 58, no. 2, pp. 629–646, 2011.

[10] J. L. Meza, V. Santibanez, R. Soto, and M. A. Llama, "Fuzzy self-tuning PID semiglobal regulator for robot manipulators," IEEE Trans. Ind. Electron., vol. 59, no. 6, pp. 2709–2717, 2012.

[11] M. J. Prabu, P. Poongodi, K. Premkumar, "Fuzzy supervised online coactive neuro-fuzzy inference system-based rotor position control of brushless DC motor," IET Power Electronics, vol. 9, no. 11, pp. 2229–2239, 2016.

[12] M. Nikkhah, H. Ashrafiuon, and F. Fahimi, "Robust control of underactuated bipeds using sliding modes," Robotica, vol. 25, no. 3, pp. 367–374, 2007.

[13] B. Xian, D. M. Dawson, M. S. de Queiroz, J. Chen, "A continuous asymptotic tracking control strategy for uncertain nonlinear systems," IEEE Trans. Autom. Control, vol. 49, no. 7, pp. 1206–1211, 2004.

[14] Z. Liu, L. Wang, C. Philip Chen, X. Zeng, Y. Zhang, and Y. Wang, "Energy-efficiency-based gait control system architecture and algorithm for biped robots," IEEE Trans. Syst., Man, Cybern. C, Appl. Rev., vol. 42, no. 6, pp. 926–933, 2012.

[15] J. Yao, Z. Jiao, and D. Ma, "Adaptive robust control of DC motors with extended state observer," IEEE Trans. Ind. Electron., vol. 61, no. 7, pp. 3630–3637, 2014.

[16] S. Kim and J. B. Bae, "Force-mode control of rotary series elastic actuators in a lower extremity exoskeleton using model-inverse time delay control (MiTDC)," IEEE/ASME Trans. Mechatronics, vol. 22, pp. 1392–1400, 2017.

[17] W. Zhang, J. B. Bae, and M. Tomizuka, "Modified preview control for a wireless tracking control system with packet loss," IEEE/ASME Trans. Mechatronics, vol. 20, pp. 299–307, 2015.

[18] J. Baek, M. Jin, and S. Han, "A new adaptive sliding mode control scheme for application to robot manipulators," IEEE Trans. Ind. Electron., vol. 63, no. 6, pp. 3628–3637, 2016.

[19] Y. Shtessel, M. Taleb, and F. Plestan, "A novel adaptive-gain supertwisting sliding mode controller: Methodology and application," Automatica, vol. 48, pp. 759–769, 2012.

[20] C. Xia, G. Jiang, W. Chen, and T. Shi, "Switch-gain adaptation current control for brushless DC motors," IEEE Trans. Ind. Electron., vol. 63, no. 4, pp. 2044–2052, 2016.

[21] M. Jin, J. Lee, and N. G. Tsagarakis, "Model-free robust adaptive control of humanoid robots with flexible joints," IEEE Trans. Ind. Electron., vol. 64, no. 2, pp. 1706–1715, 2017.

[22] D. X. Ba and J. B. Bae, "A Nonlinear Sliding Mode Controller of Serial Robot Manipulators with Two-level Gain-learning Ability," IEEE Access, 2020.

[23] D. X. Ba, "A Fast Adaptive Time-delay-estimation Sliding Mode Control for Robot Manipulators," Advances in Science, Technology and Engineering Systems Journal, Vol. 5, No. 6, 904–911, 2020.

[24] M. H. Khooban, T. Niknam, F. Blaabjerg, and M. Dehghani, "Free chattering hybrid sliding mode control for a class of nonlinear systems: electric vehicles as a case study," IET Sci. Meas. Technol, vol. 10, no. 7, pp. 776–785, 2016.

[25] H. Khooban, N. Vafamand, T. Niknam, T. Dragicevic, and F. Blaabjerg, "Model-predictive control based on Takagi-Sugeno fuzzy model for electrical vehicles delayed model," IET Electr.. Power Appl., vol. 11, no. 5, pp.918–934, 2017.

[26] J. Y. Lee, M. Jin, and P. H. Chang, "Variable PID gain tuning method using backstepping control with time-delay estimation and nonlinear damping," IEEE Trans. Ind. Electron., vol. 14, no. 12, pp. 6975–6985, 2014.

[27] D. X. Ba, H. Yeom, J. Kim, and J. B. Bae, "Gain-adaptive Robust Backstepping Position Control of a BLDC Motor System," IEEE/ASME Trans. on Mechatronics, vol. 23, no. 5, pp. 2470–2481, 2018.

[28] B. Yao, F. Bu, J. Reedy, and G. T. C. Chiu, "Adaptive robust motion control of single-rod hydraulic actuators: theory and experiments," IEEE/ASME Trans. Mechatronics, vol. 5, no. 1, pp. 79–91, Mar. 2000.

[29] K. J. Astrom and B. Wittenmark, Adaptive Control. New York: AddisonWesley, 1995.

[30] J. Yao, Z. Jiao, D. Ma, and L. Yan, "High-accuracy tracking control of hydraulic rotary actuators with modeling uncertainties," IEEE/ASME Trans. Mechatronics, vol. 19, no. 2, pp. 633–641, 2014.

[31] K. Guo, J. Wei, J. Fang, R. Feng, and X. Wang, "Position tracking control of electro-hydraulic single-rod actuator based on an extended disturbance observer," Mechatronics, vol. 27, pp. 47–56, 2015.

[32] D. X. Ba, K. K. Ahn, D. Q. Truong, and H. G. Park, "Integrated model-based backstepping control for an electro-hydraulic system," Int. J. Prec. Eng. Manufacturing, vol. 17, no. 5, pp. 1–13, 2016.

[33] M. Raibert, "Legged robots that balance," MIT Press, Cambridge, USA, 1986.

[34] D. N. Nenchev, A. Konno, and T. Tsujita, "Humanoid robots: modeling and control," Butterworth-Heinemann, Cambridge, USA, 2019.

[35] J. Lima, J. Goncalves, P. Costa, and A. P. Moreira, "Humanoid low-level control development based on a realistic simulation," Int. J. Humanoid Robotics, vol. 7, no. 4, pp. 587–607, 2010.

[36] V. M. F. Santos and F. M. T. Silva, "Design and low-level control of a humanoid robot using a distributed architecture approach," J. Vibration and Control, vol. 12, no. 12, pp. 1431–1456, 2006.

[37] F. L. Moro and L. Sentis, "Whole-body control of humanoid robot," in Humanoid Robotics: A reference, Springer, Dordrecht, 2019.

[38] J. Ahn, J. Lee, and L. Sentis, "Data-efficient and safe learning for humanoid locomotion aided by a dynamic balancing model," IEEE Robotics and Automation Letters, vol. 5, no. 3, pp. 4376–4383, 2020.

[39] J. Wang, W. Qin, and L. Sun, "Terrain adaptive walking of biped neuromuscular virtual human using deep reinforcement learning," IEEE Access, vol. 7, pp. 92465–92475, 2019.

Chapter 4

Optimal Trajectory Generation of Parallel Manipulator

Chandan Choubey and Jyoti Ohri

Abstract

In this paper we have designed an optimal trajectory generation (OTG) method to generate easy and errortless continuous path motion with quick converging by using Gray Wolf Optimization (GWO) method. This OTG method finds the trajectory path with minimum tracking-error, combined speed, joint increasing speed wrinkle as well as joint lurching move to follow a smooth along with error-free continuous path.

Keywords: Gray Wolf Optimization, Trajectory Generation algorithm, Parallel Manipulators, Error-Free Path Motion

1. Introduction

We know that the biological system basically influence the field of robot like the upper portion of human arm with a few sequential connections, this serial structure are called serial robots. The serial robots structure acquire large space, lack of precision and low load handling capability are its major drawbacks. Parallel manipulators was introduced by the researchers to reduce the disadvantages of series manipulators [1, 2].

The performance of parallel manipulator has become more advance in the recent scenario as compared to the series manipulator. The parallel manipulator has so many benefits as compared to series like precision, load handling ability, accuracy and many more. The parallel manipulators are used in aerodynamics [3], medical surgery [4–7], machine equipment's [8, 9], and object pick and place [10, 11].

A parallel manipulator consist of a movable plate connected with a fixed plate with hinged legs which is controlled by a dc motor separately. The number of orientation of legs is considered as degrees of freedom (DOF) of movable plate with respect to fixed plate, this type of arrangement is called as coupling systems.

The parallel robotic arm generally provides high and smooth speed, acceleration with accurate path tracking. For tracking continuous path, parallel manipulator must satisfied following specifications like error-free tracking, minimum settling time and robustness against uncertainties [12].

2. Literature survey

Almas shintemirov *et.al* [13] proposed an optimal path by spherical parallel manipulator (SPM), controlled by servomotors with default setting of position

IntechOpen

control. For such type of system there are three approaches to get forward kinematics, a structural space and servomotors reference paths.

Damien six *et.al* [14] had evolved a new flying robot considered as one of parallel manipulators. This robot has three similar legs connected between fixed and moving stages, are controlled by quadrotors a stage. For constructing new architecture of aerial robotics, requires quadrotor with strong body.

Soheil zarkandi [15] had proposed a parallel manipulator for CNC machine, used for object holding in the 4-axis. This manipulator has two degree-of-freedom i.e. translational and rotational. It consist of two translational DOF and one rotational DOF.

Guilherme sartori Natal *et.al* [16] had tentatively compared the R4 controlled type manipulator with three control methods and provides high acceleration. First method, the redundant controlling is specifically done by a PID controller in working space. Secondly, based on dynamic configuration of redundant R4, a dual working space with feed forward control method was implemented because of impropriate outcomes of first method. Thirdly, upgrading such type of controller, generates high acceleration above 100G in order to achieve path tracking.

Bikash kumar Sarkar [17], had shown the reproduction concentrate over the using pressurized water activated 2DOF equal controller pondered to the posture (hurl and pitch) control application. The framework model is pondered to the ease pressure driven part setups like corresponding valve with dead band, low speed water powered chamber and so forth. The streamlined numerical model of the controller has been created in this investigation. To examine the control execution by a model free fluffy tuned feed forward inclination PID regulator for present control application, this model has been utilized.

Yogesh singh *et.al* [18] has introduced a controller called U-formed planar equal controller which tends to the opposite elements of three levels of opportunity (DOF) where the introduced controller has three legs comprising of kaleidoscopic revolute (PPR) joint course of action in which every leg had one dynamic kaleidoscopic joint. A versatile sliding mode control, joined with an aggravation onlooker for the movement control of the controller was evaluated like a relative subsidiary (PD). The controlled mechanical controller was changed into decoupled elements utilizing this plan thus that the movement execution was gainful to quantify.

Jiantao Yao *et.al* [19] has built up the ability and the exhibition of the equal controller as for the repetitive incitation. While expanding a drive for the center PRPU uninvolved imperative branch to make it an excess incitation branch, it is expected to manage coordination and circulation of the main impetus of the equal controller with repetitive activation and need to understand the control system dependent on elements, based on the first 5UPS/PRPU equal controller. The component that exist the repetitive incitation from the viewpoint of level of opportunity and settled in a powerful model dependent on Lagrangian technique is improved by bringing in the arrangements excess sorts and pieces of 5UPS/PRPU equal controller with repetitive activation.

3. Optimal trajectory generation algorithm

The industrial robots were broadly used in various fields like automotive and aircraft industries and many more. The use of industrial robots, generally carry out repeated tasks such as pick and place, welding, assembling, etc. Their adaptability and capability to perform complex tasks in a significant workspace makes them useful in SME (small and medium enterprise). The characteristic advantages they offer in machine applications like prototyping, cleaning and pre-machining of cast parts as well as in end-machining of middle tolerance parts, have increased their usage rapidly.

An input is produced in the control system of the robotic manipulators by trajectory generation for executing the required task with satisfactory performance since a path-constrained motion is followed by the robotic manipulator. The path of the robotic trajectories is assembled offline at first, and later it is assembled online by the end-effectors. There are two approaches in offline trajectory - hand level and joint level. By using Jacobian transformation, these joint coordinates are transformed into Cartesian coordinates for each sampling.

By using opposite kinematics, the Cartesian coordinates are transformed into joint coordinates. Joint level approach costs less expensive in terms of computational complexity than other approaches while controlling the robotic manipulators. Moreover, this joint level approach has an added advantage of considering only the kinematic constraints during the trajectory generation, while ignoring the dynamic constraints that increase the computational effort.

An Optimal Trajectory Generation Algorithm (OTGA) [20] is developed to generate smooth motion trajectories with minimum time for Dof parallel manipulators. For optimal trajectory generation, the Gray Wolf technique is employed with constraints and objective functions, this proposed OTG algorithm uses minimal tracking error. Moreover, for smooth continuous motion of the robotic manipulators, joint speed, acceleration and jerks were also considered along with it. So by using both objective constraints, the Gray Wolf optimization technique selects an optimal trajectory at every iteration as shown in **Figure 1**.

3.1 Trajectory generation remarks

A reference trajectory is created by using the developed Optimal Trajectory Generation [21] manipulators. The path constraint motion of the industrial robots plays a vital role in welding, cutting, surgery and machining applications. A sample reference trajectory with 15 segments is shown below in **Figure 2**.

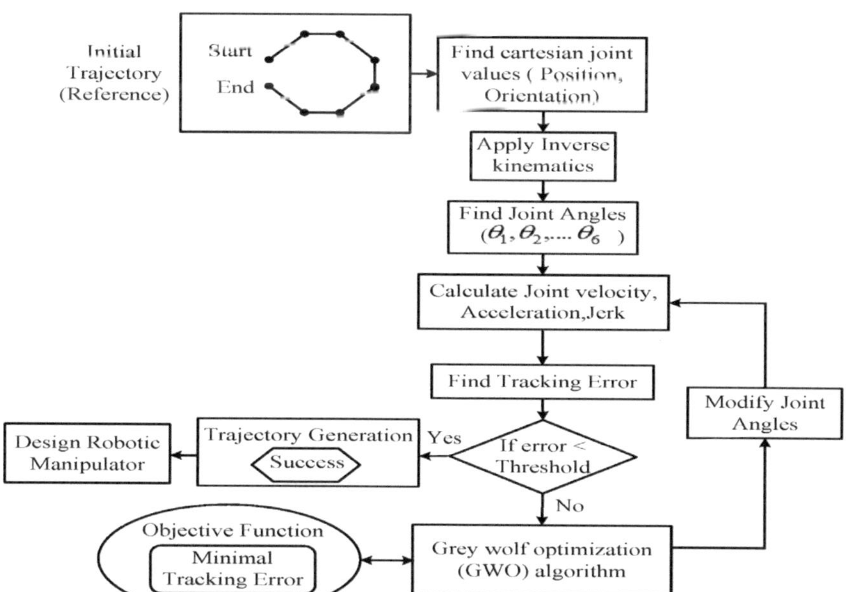

Figure 1.
The schematic diagram of OTGA.

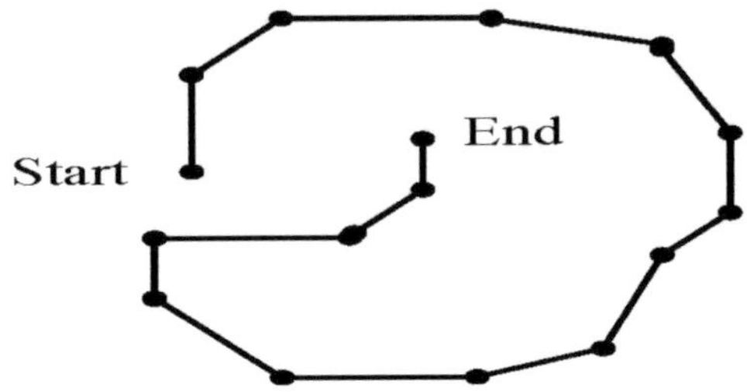

Figure 2.
Reference trajectory.

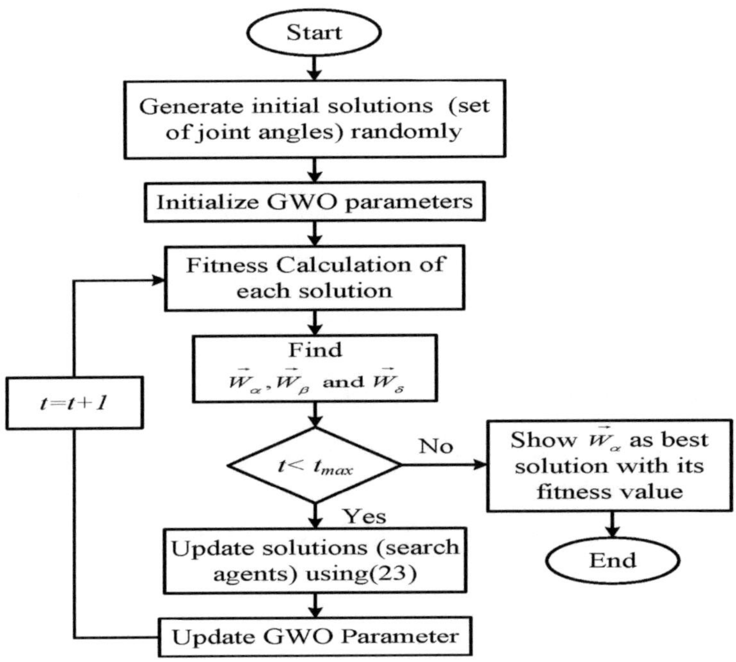

Figure 3.
GWO flow chart.

Based on the reference trajectory, the design of 6 DOF robotic manipulator is analyzed. For analysis, primary trajectory is approximately created and then optimized using Gray Wolf optimization algorithm. The primary trajectory is calculated for each segment starting from the 'Start' segment to the 'End' segment on the reference trajectory.

3.2 GWO for the optimal trajectory generation

In this section, the Gray Wolf Optimization (GWO) algorithm is employed for the optimal trajectory generation [21–26]. Here, initially the joint coordinates of the parallel robotic arms are obtained using opposite kinematic approach. Then this sets of joint coordinates, they are optimally fixed by minimizing the path tracing error. After this, the manipulator joint coordinates like speed, acceleration and jerk are calculated by utilizing the finest set if joint angles. The flowchart for the GWO methodology is described below in **Figure 3**.

3.2.1 Objective function

For selecting and tracking of a continuous path with minimal tracking error is the prime function of the developed OTG method. This tracking error function can be framed as,

$$Tracking\ Error = \min\left(p_d(i) - p_a(i)\right) \tag{1}$$

Where, $p_d(i)$ denotes the desired trajectory for $(i)^{th}$ robotic arm; $p_a(i)$defines the current trajectory for $(i)^{th}$ robotic arm. Also, the following conditions must be satisfied in order to get a smooth continuous path motion.

$$\min\left\{J_v, J_a, J_j\right\} \tag{2}$$

Such as, $J_v = \frac{d\theta}{dt}; J_a = \frac{dJ_v}{dt}; J_j = \frac{dJ_a}{dt}$.

On Basis of minimum tracking error, during each recurrence, joint velocity (j_v), joint acceleration (Ja) and joint jerk (J_j) obtained for each set of manipulator joint.

3.3 The Gray Wolf Optimization (GWO)

The Gray Wolf Optimization (GWO) is an effective optimization method which emulates the leadership grading, trapping protocol and hunting mechanism of gray wolves in nature. In GWO, the process of optimal trajectory tracking is performed by four gray wolves namely; alpha, beta, delta and omega wolves. Decisions about hunting, time and place are being made by the alpha, The beta and gamma wolves is basically considered as subordinate wolves that help the alpha in decision making, the timid part of the gray wolves hierarchy is being represented by omega only.

3.3.1 Fitness evaluation

Following steps are there to evaluate the fitness function as written below.

$$Fitness\left(A_n^t\right) = \min\left(Tracking\ Error\right) + \min\left\{J_v, J_a, J_j\right\} \tag{3}$$

Mathematical approach for search operation:

In GWO, the α, β and δ wolves guide each other and encircles the prey. It is pretended that α, β and δ gives an appropriate knowledge regarding the exact location of the prey from overall solution. Due to which, the primary best solutions are achieved and now it is considered to generate newer solutions, which can be systematically established as beneath:

$$\vec{W}(t+1) = \vec{W}_r(t) - \vec{M} \cdot \vec{Q} \tag{4}$$

In the above Eq. (4), $\vec{Q}$can be given as,

$$\vec{Q} = \left|\vec{S} \cdot \vec{W}_r(t) - \vec{W}(t)\right| \tag{5}$$

$\vec{W}$ is represented as the gray wolf actual location, $\vec{W}_r$ represents prey desired location, $\vec{M}\ and\ \vec{S}$represents the coefficient vectors respectively and 't' denotes the no.

of operations. We can obtain the coefficient vectors $\vec{M}$ *and* $\vec{S}$ by the equations given below:

$$\vec{M} = 2\vec{m} \cdot \vec{x}_1 - \vec{m} \tag{6}$$

$$\vec{S} = 2 \cdot \vec{x}_2 \tag{7}$$

Here '$\vec{m}$' denotes the constant value that decreases from 2 to 0 and $\vec{x}_1$ and $\vec{x}_2$ denotes any random values between $[0, 1]$. $\vec{m}$ is selected within the range between 2 to 0 in every operations as per the below equation,

$$\vec{m} = 2 - t\left(\frac{2}{t_{\max}}\right) \tag{8}$$

Where, '$t_{\max}$' represents the maximum allowed iterations. Assuming that, the information about the position of prey is possibly confirmed by the Alpha, Beta and Delta solutions; whereas the updates in position of Omegas is govern by previous solutions. The position updating of wolves is depended on all three best solutions as shown below:

$$\vec{W}_1 = \left|\vec{W}_\alpha(t) - \vec{M}_1 \cdot \vec{Q}_\alpha\right| \tag{9}$$

$$\vec{W}_2 = \left|\vec{W}_\beta(t) - \vec{M}_2 \cdot \vec{Q}_\beta\right| \tag{10}$$

$$\vec{W}_3 = \left|\vec{W}_\delta(t) - \vec{M}_3 \cdot \vec{Q}_\delta\right| \tag{11}$$

Where, $\vec{Q}_\alpha$, $\vec{Q}_\beta$, $\vec{Q}_\delta$ are calculated as:

$$\vec{Q}_\alpha = \left|\vec{S}_1 \cdot \vec{W}_\alpha - \vec{W}\right| \tag{12}$$

$$\vec{Q}_\beta = \left|\vec{S}_2 \cdot \vec{W}_\beta - \vec{W}\right| \tag{13}$$

$$\vec{Q}_\delta = \left|\vec{S}_3 \cdot \vec{W}_\delta - \vec{W}\right| \tag{14}$$

Based on the above Eqs. (10)-(12), the solution for next iteration will be obtained as follows:

$$\vec{W}(t+1) = \frac{\left(\vec{W}_1 + \vec{W}_2 + \vec{W}_3\right)}{3} \tag{15}$$

The process of updating of wolf current positions takes place continuously until the maximum iteration is reached. If the overall optimum solution is does not reached to its maximum, or likewise the new solution will be updated for which the best feasible solution take place and hence based on the best suitable solution the next updates will be executed continuously. Due to this, the optimal continuous path is selected with error-free tracking path.

3.4 The Genetic Algorithm (GA)

Genetic Algorithm (GA) is ordered among three distinct parts for example multiplication, hybrid and change and it is expounded momentarily in couple of

Algorithm Parameters	Outcome
Variables counts	6 $[\theta_1, \theta_2, \theta_3, \theta_4, \theta_5, \theta_6]$
Maximum Generation	250
Population Size	60
Encoding	Binary
Selection	Uniform
Crossover	0.7
Mutation	0.3
Total number of counts	258

Table 1.
Transformative algorithm parameters of GA.

steps [27]. The chromosome shaped by six factors of lattice $\theta(jointangle)$ [28, 29] to accomplish ideal worth. Variation boundaries of GA are appeared in the beneath **Table 1**.

4. Result analysis

In this section, the analysis for the developed GWO based OTG method and GA for optimal planning of the trajectory for designing the 6 DOF Robotic manipulator. The applied methods are implemented by MATLAB.

A 3-degree-of-freedom (3-DOF) planar parallel manipulator performing high-speed, high-acceleration, and high-accuracy trajectory tracking as similar to the novel experimental pick-and-place manipulator is designed and constructed. At the time of trajectory tracking, multiple closed-loop performance specifications like tracking accuracy, settling time, control effort, and robustness to parameter uncertainty must be satisfied simultaneously. Commonly, closed loop requirement is clashing, i.e., when one requirement is improved, others may break down.

An Optimal Trajectory Generation Algorithm (OTGA) is created for producing least time smooth movement directions for 6 DOF equal controllers. The proposed OTGA utilizes the Gray Wolf enhancement procedure for the ideal direction age utilizing numerous goal capacities. Alongside this, to follow the smooth movement of mechanical controllers, the joint speed, joint increasing speed and joint jerks requires optimal value. At each cycle, the proposed Gray Wolf improvement method chooses the ideal directions utilizing the goal limitations.

The below graph 1 to 6 in **Figure 4** shows, the comparison of joint velocity for all active joints angles $\theta_1, \theta_2, \ldots, \theta_6$ of the manipulator between the proposed and existing methods. The below simulated results shows smooth motion with optimal velocity at each joints of the robotic arm (manipulator).

The below graph 1 to 6 in **Figure 5** shows, the comparison of joint acceleration for all active joints angles $\theta_1, \theta_2, \ldots, \theta_6$ of the manipulator between the proposed and existing methods. The below simulated results shows smooth motion with optimal acceleration at each joints of the robotic arm (manipulator).

The below graph 1 to 6 in **Figure 6** shows, the comparison of joint jerks for all active joints angles $\theta_1, \theta_2, \ldots, \theta_6$ of the manipulator between the proposed and existing methods. The below simulated results shows smooth motion with minimum jerks at each joints of the robotic arm (manipulator).

The above figures show that the comparison between projected GWO technique, existing GA and default methods for trajectory generation. We have taken

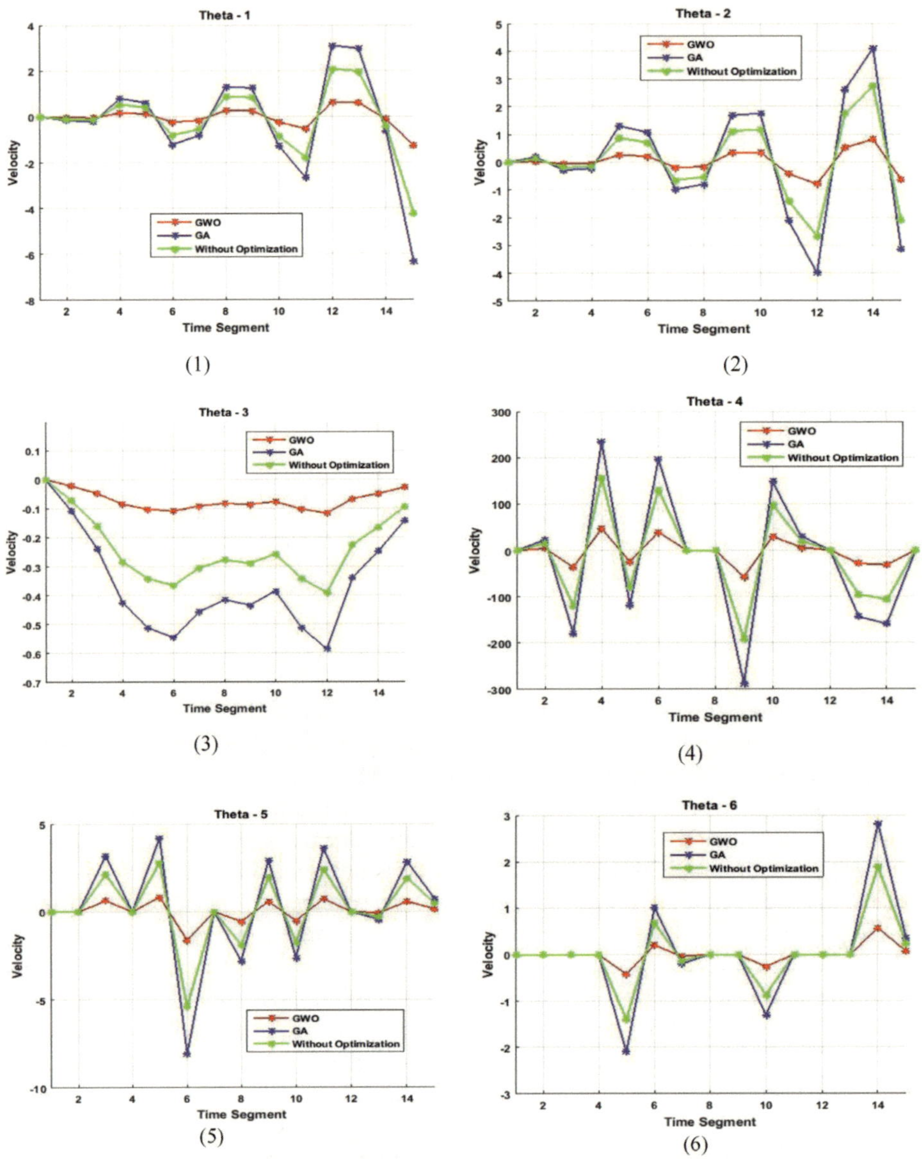

Figure 4.
Comparison plot for time segment vs. velocity for proposed and existing method.

three measurements named as acceleration, jerk and velocity in which all these are going to be compare with different time segments. It is clearly shows that the proposed method achieves minimum effective value as compared to the exiting techniques.

5. Discussion and conclusions

An optimal trajectory generation methodology is proposed which generates errorless continuous path motion with fast converging the Gray Wolf Optimization (GWO) method. The proposed OTG method using GWO algorithm is compared

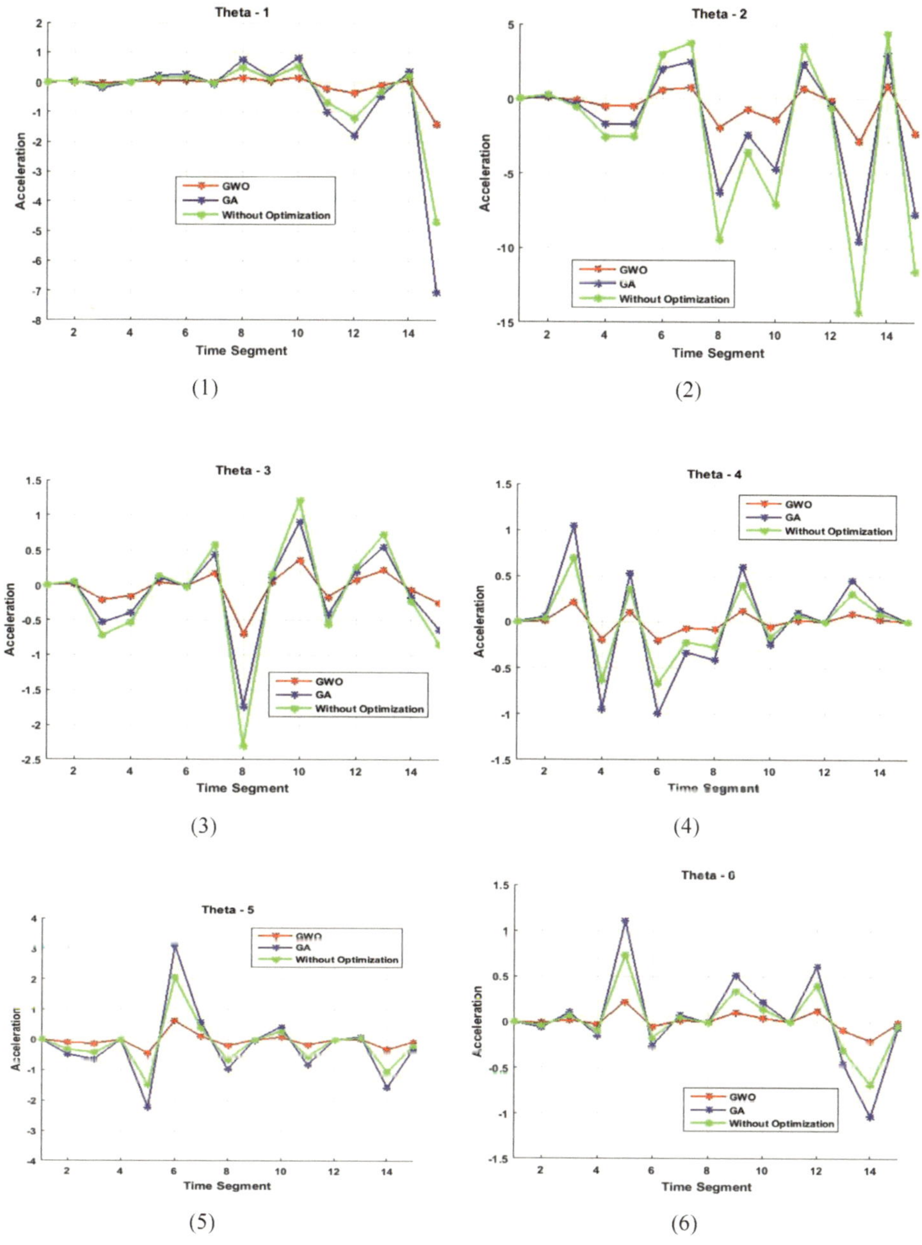

Figure 5.
Comparison plot for time segment vs. acceleration for projected and existing method.

with the GA (Genetic Algorithm) based trajectory generation method and a traditional trajectory generation method.

The mean, maximum and minimum acceleration value is also less for the proposed OTG with GWO method when compared to the existing methods. The least acceleration value is attained for the joint angle. Finally, the Joint jerk value is also calculated for all the joint angles using proposed and exiting methods with 15 segments.

The comparison of joint velocity, joint acceleration and joint jerks for all active joints angles $\theta_1, \theta_2, \dots, \theta_6$ of the manipulator between the proposed and existing

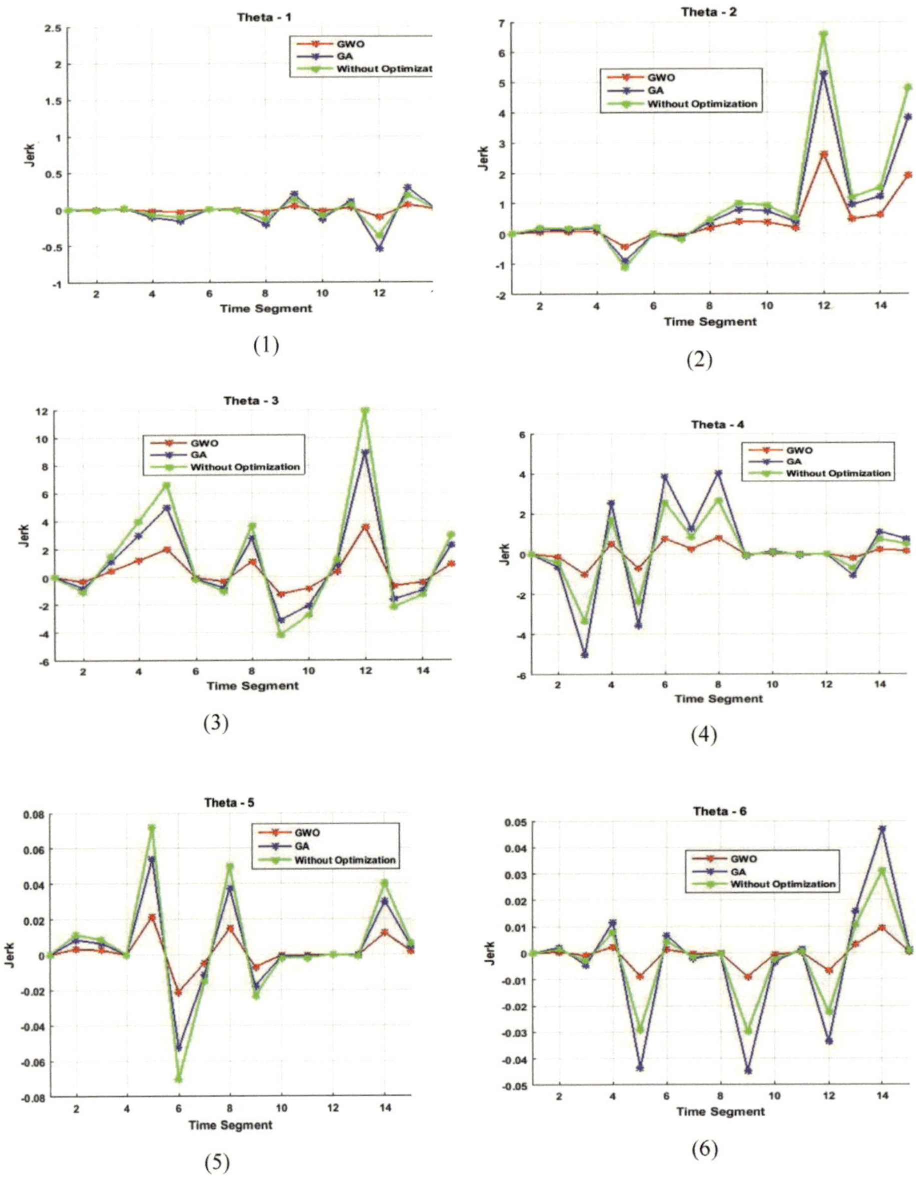

Figure 6.
Comparison plot for time segment vs. jerk for proposed and existing method.

methods. The below simulated results shows smooth motion with optimal velocity at each joints of the robotic arm (manipulator).

Comparison results can be summarized as follows:

i. The maximum average velocity of the proposed GWO based OTG is observed 1.75 times lesser than GA based OTG and 1.01 times greater than non-optimize method.

ii. The acceleration maximum average value of the proposed GWO based OTG is observed that 4.03 times and 3.92 times lesser than GA based OTG and non-optimize method.

iii. The jerk maximum average value of the proposed GWO based OTG is observed that 2.41 times and 2.04 times lesser than GA based OTG and non-optimize method.

iv. Proposed OTG GWO generates minimum 118.4% and maximum 236.1% better velocity, minimum 156.4% and maximum 592% better acceleration, and minimum 108.7% and maximum 310.7% better jerk.

The efficiency of projected methodology has been analyzed with the actual research works. The experimental result shows that a good optimization of developed OTG method in terms of shared speed, joint speed ripples, and joint lurching move measures. This proves that the proposed OTG algorithm works effectively to follow the optimal trajectory with less tracking error and smooth continuous path motion.

Author details

Chandan Choubey[1]* and Jyoti Ohri[2]

1 Department of Electronics and Communication Engineering, Dronacharya Group of Institutions, Greater Noida, India

2 Department of Electrical Engineering, National Institute of Technology, Kurukshetra, India

*Address all correspondence to: er.choubey.chandan@gmail.com

References

[1] Mehta, Vivek Kumar, and Bhaskar Dasgupta. "A general approach for optimal kinematic design of 6-DOF parallel manipulators." *Sadhana* 36, no. 6 (2011): 977-994.

[2] Wu, Xiuming, Tomohiro Kobayashi, Akira Nakamura, Kenichiro Yasui, and Hideo Furuhashi. "Development of the upper body of a humanoid robot using parallel linkage mechanisms." In *Industrial Automation, Information and Communications Technology (IAICT), 2014 International Conference on*, pp. 9-14. IEEE, 2014.

[3] L.W. Tsai, Robot Analysis, "The Mechanics of Serial and Parallel Manipulators. Hoboken," NJ, USA: Wiley, pp. 134–142, 1999.

[4] Stewart D, "A platform with six degrees of freedom," Proceedings of the Institution of Mechanical Engineers. pp. 371-86, 1965.

[5] Y. Li and Q. Xu, "Design and development of a medical parallel robot for cardiopulmonary resuscitation," IEEE Trans. Mechatronics, vol. 12, no. 3, pp. 265–273, Jun. 2007.

[6] M. Shoham, M. Burman, E. Zehavi, L. Joskowicz, E. Batkilin, and Y. Kunicher, "Bone-mounted miniature robot for surgical procedures: Concept and clinical applications," IEEE Trans. Robot. Autom., vol.19, no.5, pp. 893–901, Oct. 2003.

[7] W. L. Xu, J. S. Pap, and J. Bronlund, "Design of a biologically inspired parallel robot for foods chewing," IEEE Trans. Ind. Electron., vol. 55, no. 2, pp. 832–841, Feb. 2008.

[8] W. L. Xu, J. D. Torrance, B. Q. Chen, J. Potgieter, J. E. Bronlund, and J. S. Pap, "Kinematics and experiments of a life-sized masticatory robot for characterizing food texture," IEEE Trans. Ind. Electron., vol. 55, no. 5, pp. 2121–2132, May 2008.

[9] E. Ottaviano and M. Ceccarelli, "Application of a 3-DOF parallel manipulator for earthquake simulations," IEEE Trans. Mechatronics, vol. 11, no. 2, pp. 241–246, Apr. 2006.

[10] F. J. Berenguer and F. M. M. Huelin, "Zappa, a quasipassive biped walking robot with a tail: Modeling, behaviour and kinematic estimation using accelerometers," IEEE Trans. Ind. Electron., vol. 55, no. 9, pp. 3281– 3289, Sep. 2008.

[11] F. Pierrot, V. Nabat, O. Company, S. Krut, and P. Poignet, "Optimal design of a 4-DOF parallel manipulator: From academia to industry," IEEE Trans. Robot., vol. 25, no. 2, pp. 213–224, Apr. 2009.

[12] Ren, Lu, James K. Mills, and Dong Sun. "Trajectory tracking control for a 3-DOF planar parallel manipulator using the convex synchronized control method." *IEEE Transactions on Control Systems Technology* 16, no. 4 (2008): 613-623.

[13] Almas Shintemirov, Aibek Niyetkaliyev, and Matteo Rubagotti. "Numerical optimal control of a spherical parallel manipulator based on unique kinematic solutions." *IEEE/ ASME Transactions on Mechatronics* 21, no. 1 (2016): 98-109.

[14] Damien Six, Sébastien Briot, Abdelhamid Chriette, and Philippe Martinet. "The Kinematics, Dynamics and Control of a Flying Parallel Robot with Three Quad rotors." *IEEE Robotics and Automation Letters* 3, no. 1 (2018): 559-566.

[15] Soheil Zarkandi. "Kinematic and dynamic modeling of a planar parallel manipulator served as CNC tool holder."

International Journal of Dynamics and Control 6, no. 1 (2018): 14-28.

[16] Guilherme Sartori Natal, Ahmed Chemori, and François Pierrot. "Dual-space control of extremely fast parallel manipulators: Payload changes and the 100g experiment." *IEEE Transactions on Control Systems Technology* 23, no. 4 (2015): 1520-1535.

[17] Bikash Kumar Sarkar. "Modeling and validation of a 2-DOF parallel manipulator for pose control application." *Robotics and Computer-Integrated Manufacturing* 50 (2018): 234-241.

[18] Yogesh Singh, V. Vinoth, Y. Ravi Kiran, Jayant Kumar Mohanta, and Santha kumar Mohan. "Inverse dynamics and control of a 3-DOF planar parallel (U-shaped 3-PPR) manipulator." *Robotics and Computer-Integrated Manufacturing* 34 (2015): 164-179.

[19] Jiantao Yao, Weidong Gu, Zongqiang Feng, Lipu Chen, Yundou Xu, and Yongsheng Zhao. "Dynamic analysis and driving force optimization of a 5-DOF parallel manipulator with redundant actuation." *Robotics and Computer-Integrated Manufacturing* 48 (2017): 51-58.

[20] Choubey, C., & Ohri, J., "Optimal Trajectory Generation for a 6-DOF Parallel Manipulator Using Grey Wolf Optimization Algorithm," *Robotica*, pp.1-17, 2020.

[21] A. Ahmed, R. Gupta and G. Parmar, "GWO/PID Approach for Optimal Control of DC Motor," *5th International Conference on Signal Processing and Integrated Networks (SPIN)*, Noida, pp. 181-186, , 2018.

[22] Tripathi S., Shrivastava A., Jana K.C, "GWO Based PID Controller Optimization for Robotic Manipulator," *Intelligent Computing Techniques for Smart Energy Systems. Lecture Notes in Electrical Engineering*, vol.607, 2020.

[23] B. McAvoy, B. Sangolola and Z. Szabad, "Optimal trajectory generation for redundant planar manipulators," *IEEE international conference on systems, man and cybernetics. 'cybernetics evolving to systems, humans, organizations, and their complex interactions'*, pp. 3241-3246 vol.5, 2000.

[24] Serdar Kucuk, "Maximal dexterous trajectory generation and cubic spline optimization for fully planar parallel manipulators," *Computers & Electrical Engineering*, pp 634-647, Vol. 56, 2016.

[25] N. Kubota, T. Arakawa, T. Fukuda and K. Shimojima, "Trajectory generation for redundant manipulator using virus evolutionary genetic algorithm," *Proceedings of International Conference on Robotics and Automation, Albuquerque*, NM, USA, pp. 205-210, Vol.1, 1997.

[26] Serdar Kucuk, "Optimal trajectory generation algorithm for serial and parallel manipulators," *Robotics and Computer-Integrated Manufacturing*, Vol. 48, pp. 219-232, 2017.

[27] P. S. Oliveira, L. S. Barros and L. G. de Q. Silveira Júnior,"Genetic Algorithm Applied to State Feedback Control Design," *IEEE conf. on Transmission and Distribution and Exposition*,2012.

[28] M.W. Chen, A.M.S. Zalzala, "Dynamic modelling and genetic-based trajectory generation for non-holonomic mobile manipulators," *Control Engineering Practice*, Vol. 5(1), pp. 39-48, 1997.

[29] Gaurav Chaudhary and Jyoti Ohri, "3-DOF Parallel manipulator control using PID controller", *IEEE International Conference on Power Electronics, Intelligent Control and Energy Systems*, pp.1-6, 2016.

Chapter 5

Guidance and Control of a Planar Robot Manipulator Used in an Assembly Line

Bülent Özkan

Abstract

In order to achieve higher productivity and lower cost requirements, robot manipulators have been enrolled in assembling processes in last decades as well as other implementation areas such as transportation, welding, mounting, and quality control. As a new application of this field, the control of the synchronous movements of a planar robot manipulator and moving belt is dealt with in this study. Here, the mentioned synchronization is tried to be maintained in accordance with a guidance law which leads the robot manipulator to put selected components onto the specific slots on the moving belt without interrupting the assembling process. In this scheme, the control of the manipulator is carried out by considering the PI (proportional plus integral) control law. Having performed the relevant computer simulations based on the engagement geometry between the robot manipulator and moving belt, it is verified that the mentioned pick-and-place task can be successfully accomplished under different operating conditions.

Keywords: Guidance and control, robot manipulator, linear homing guidance law, assembly line, automation

1. Introduction

Robot manipulators have been utilized in many application areas since 1960's [1, 2]. In addition to their implementations in harsh and unusual environments involving tedious, hard, and hazardous tasks, the productivity, cost reduction, and time effectiveness considerations have put forward the use of the manipulators in the production and assembly applications [3–5].

Regarding the pick-and-place operations in which certain components are placed onto specific slots on a moving belt by means of the end effector of the robot manipulator that constitute the hand of the manipulator, the most common method is to make the placement of the component to the slot once they coincide. This attitude has been chosen by some famous vehicle manufacturers [6]. Since it is required to halt the moving belt at coincidences of the end effector of the robot and slot in this scheme, a discrete motion strategy is developed for this purpose. Even though this approach works well when relatively light components are under consideration, the increment in the component mass leads to higher acceleration requirements to speed up the belt right after the placement. In such a scheme with a robot manipulator, the belt should be halted at specific points in order to allow the manipulator to put the component on the slot. In fact, this means using powerful actuators which are big and expensive in practice and hence it violates the cheapness demand.

As a remedy to the preceding method, it seems reasonable to the motion of the belt even during placement. This results in diminishing the power need in operation and allows to use smaller and cheaper actuators, or motors [7]. On the other hand, it may not possible to coincide the end effector of the robot manipulator grasping the components and belt all the time due to uncertain factors such as nonlinear friction effects on the belt dynamics when larger and heavier parts are considered as in automotive industry. In order to compensate this weakness of conventional motion planning strategies based on making the placements upon the pre-calculated coincidence positions ignoring the probable uncertainties, "guidance" approach can be utilized in continuous-time engagements.

In addition to optimization-based motion planning schemes based on the minimum time and/or minimum energy expenditure criteria, a hybrid target point interception algorithm is proposed as schematized in **Figure 1** where the abbreviation AIPNG stands for the "augmented ideal proportional navigation guidance" for target catching [8–12]. In the mentioned studies in which the position information is often acquired by visual sensors, the engagement models including planar manipulators having two or three degrees of freedom, or two or three links, in general are validated through computer simulations [13, 14].

Guidance laws developed originally for the munitions against specific targets can be adapted to the motion planning of the robot manipulators that can be thought as "very short-range missiles" regarding their connections to the ground [15]. As an advantage over the munitions which have generally no thrust support during their guidance phase, the robot manipulators can be accelerated along their longitudinal axis [13]. In early applications, the robotic arms were tried to be guided by means of certain sensors placed on the end effectors such as optical sensors operating along with laser beams and visual sensors, i.e. cameras [16, 17]. As a distinguished implementation of guided robot manipulators, the guidance of micromanipulators utilized in microsurgery is accomplished by the visual guidance of the operator, i.e. surgeon [18]. The vision-based guidance approach is proposed for tele-robotic systems as well [2]. Moreover, the guidance of mobile robots is dealt with in swarm arrangements [19]. In another robotic application, the proportional navigation

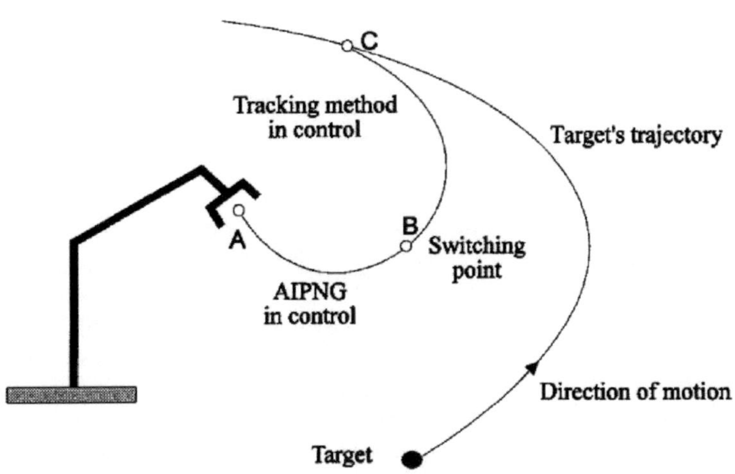

Figure 1.
Hybrid target interception scheme [8].

guidance (PNG) law which is very popular in aerial systems is considered with different navigation constants under the absence and presence of obstacles on the collision trajectory between the manipulator and target object and relative computer simulations are performed [20].

In order to synchronize the movements of the robot manipulator and moving belt in a continuous operation, it is a viable way to make them compatible in speed sense. This solution reduces to order of the robot manipulator dynamics two to one and thus most of the overshoots in the transient motion phase of the manipulator can be prevented [21].

The control strategy is very significant for the realization of commands generated by the considered guidance law. For this purpose, certain control methods are encountered in the literature for the robot manipulators such as H_2/H_∞ norm-based robust control scheme supplemented by a Takagi-Sugeno type fuzzy control such that parameter uncertainties and nonlinear effects are accounted [22]. Similar to this work, the control of a prosthetic leg is handled regarding a stable robust adaptive impedance control [23]. The adaptive control of robot manipulators has become one of the most popular research areas for last decades. The control schemes based on classical laws such as PD (proportional plus derivative) law are proposed against parameter uncertainties and unmodeled disturbances. The effectiveness of the suggested approaches is then tried to be demonstrated by well-designed computer simulations and experimental studies [24, 25].

In this study, a guidance-based motion planning approach utilizing the linear homing guidance (LHG) law is proposed for the engagement problem between a planar two-link robot manipulator shown in **Figure 2** with an origin point O and moving belt in a continuous manner [20]. Although the LHG law generates the guidance commands in terms of the linear velocity components of the tip point of the manipulator, it is more reasonable to control the manipulator through the corresponding joint variables because the actuators are connected to the joints. Therefore, an indirect adaptive control system based on the computer torque method is designed by continuously updating the controller gains during the operation after transforming the guidance commands to the joint space [13, 26]. As per the data acquired from the computer simulations conducted in the MATLAB®

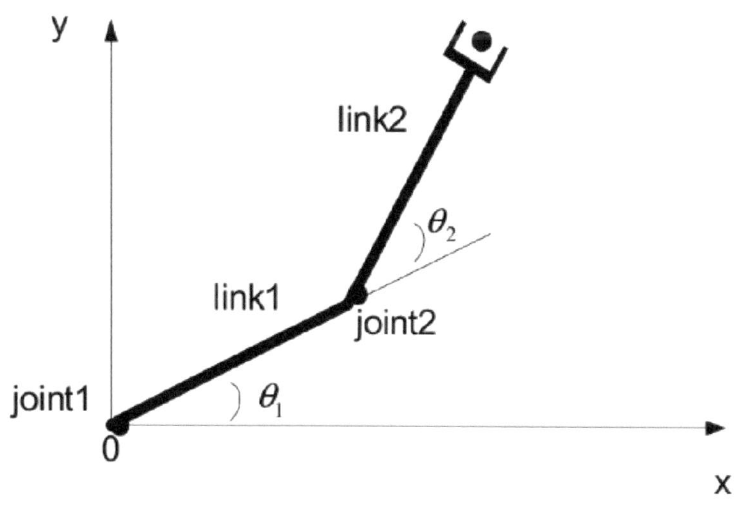

Figure 2.
Two-link robot manipulator [9].

SIMULINK® environment, it is decided that the present approach can be applied on mounting lines to attain affordable and cheaper processes.

2. System Definition

Taking the component as a two degrees of freedom spherical mass whose position is fully defined by the horizontal and lateral linear position components of a point on it, i.e. point P, at any time with no orientation, it suffices to have a robot manipulator with two degrees of freedom to move the component to any point on the horizontal plane within its kinematic limits. The schematic view of the system involving the two-link robot manipulator and moving belt is submitted in **Figure 3** along with the corresponding definitions listed below.

x and y: horizontal and lateral axes of the inertial frame represented by F_O.

$\vec{u}_1^{(0)}$ and $\vec{u}_2^{(0)}$: unit vectors denoting the x and y axes of F_O.

O and A: joints of the robot manipulator.

a_1 and a_2: lengths of the first and second links of the robot manipulator.

θ_1 and θ_2: relative rotation angles of the first and second joints of the robot manipulator.

P: point taken on the end effector of the robot manipulator.

x_P and y_P: horizontal and lateral position components of point P.

S: mid-point of the slot on the moving belt.

S_i: changing points of the shape of the moving belt (i = 1, 2, 3, and 4).

v_S: speed of the slot on the moving belt.

x_S and y_S: horizontal and lateral position components of point S.

ρ: turn radius of the moving belt.

ψ: rotation angles of the moving belt on its circular tip portions.

L: total length of the moving belt.

d: perpendicular distance between the connection point of the robot manipulator to the ground and the center line of the portion of the moving belt in the closest position to that point.

$\vec{g}$: gravity vector (g = 9.81 m/s^2).

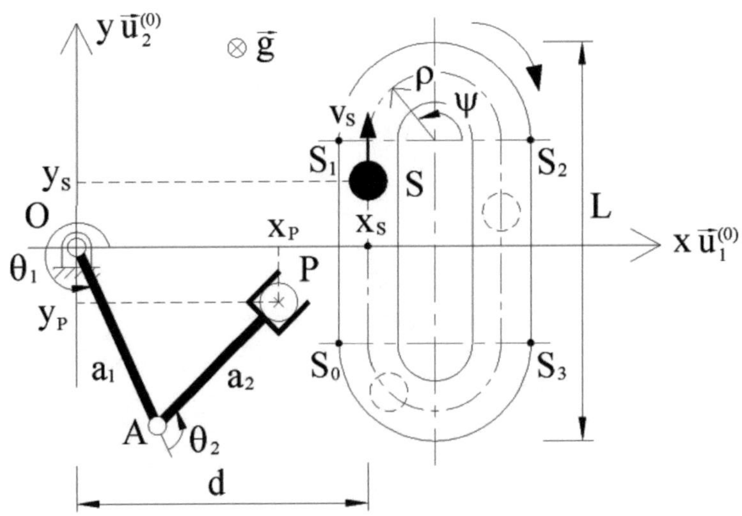

Figure 3.
System of the robot manipulator and moving belt.

3. Robot Manipulator Kinematics

In order to transform the guidance commands to the linear velocity components of the manipulator tip point into the angular speed variables of the joints, the kinematic relationships among those variables are considered.

Thus, the column vector of the position components of point P on the end effector ($\bar{r}_P$) can be written in terms of θ_1 andθ_2 in the next manner:

$$\bar{r}_P = a_1 e^{j\theta_1} + a_2 e^{j\theta_{12}} \tag{1}$$

where $\bar{r}_P = \begin{bmatrix} x_P & y_P \end{bmatrix}^T$ and $\theta_{12} = \theta_1 + \theta_2$ with $j = \sqrt{-1}$ while the letter "e" stands for the "exponential" operation.

Resolving Eq. (1) into its components, the equations given below come into the picture:

$$x_P = a_1 \cos(\theta_1) + a_2 \cos(\theta_{12}) \tag{2}$$

$$y_P = a_1 \sin(\theta_1) + a_2 \sin(\theta_{12}) \tag{3}$$

In velocity level, the following matrix expression is found by taking the time derivative of Eq. (1) with $\bar{\theta} = [\theta_1 \quad \theta_2]^T$:

$$\dot{\bar{r}}_P = \hat{J}_P \dot{\bar{\theta}} \tag{4}$$

In Eq. (4), the Jacobian matrix of the manipulator tip point is defined as follows:

$$\hat{J}_P = \begin{bmatrix} -a_1 \sin(\theta_1) - a_2 \sin(\theta_{12}) & -a_2 \sin(\theta_{12}) \\ a_1 \cos(\theta_1) + a_2 \cos(\theta_{12}) & a_2 \cos(\theta_{12}) \end{bmatrix} \tag{5}$$

From Eq. (4), the angular velocities of the manipulator links can be obtained as given below:

$$\dot{\bar{\theta}} = \hat{J}_P^{-1} \dot{\bar{r}}_P \tag{6}$$

The "elbow-up" configuration of the manipulator in which the joint indicated by letter A in **Figure 3** becomes in the upper position is taken into account in the inverse kinematic calculation above.

Eventually, the linear acceleration equations come into the picture by taking the time derivative of Eq. (6):

$$\ddot{\bar{\theta}} = \hat{J}_P^{-1} \left(\ddot{\bar{r}}_P - \dot{\hat{J}}_P \dot{\bar{\theta}} \right) \tag{7}$$

In Eq. (7), the time derivative of the tip point Jacobian matrix are determined in the forthcoming fashion with $\dot{\theta}_{12} = \dot{\theta}_1 + \dot{\theta}_2$:

$$\dot{\hat{J}}_P = \begin{bmatrix} -a_1 \dot{\theta}_1 \cos(\theta_1) - a_2 \dot{\theta}_{12} \cos(\theta_{12}) & -a_2 \dot{\theta}_{12} \cos(\theta_{12}) \\ -a_1 \dot{\theta}_1 \sin(\theta_1) - a_2 \dot{\theta}_{12} \sin(\theta_{12}) & -a_2 \dot{\theta}_{12} \sin(\theta_{12}) \end{bmatrix} \tag{8}$$

4. Dynamic Modeling of the Robot Manipulator

The governing differential equations of motion of the robot manipulator schematized in **Figure 3** can be derived using the well-known virtual work method [9].

Neglecting the gravity vector ($\vec{g}$) for the present engagement described on the horizontal plane, the equations of motion of the manipulator can be derived in the matrix form as follows:

$$\overline{T} = \hat{M}(\overline{\theta})\,\ddot{\overline{\theta}} + \hat{H}\left(\dot{\overline{\theta}},\overline{\theta}\right)\dot{\overline{\theta}} \tag{9}$$

In Eq. (9),$\overline{T} = \begin{bmatrix} T_1 & T_2 \end{bmatrix}^T$ stands for the torque column matrix with T_1 and T_2 which denote the control torques applied to the first and second joints of the robot manipulator, respectively. Here, the superscript "T" indicates the "transpose" operation. Also, the inertia and compound friction and Coriolis matrices [$\hat{M}(\overline{\theta})$ and $\hat{H}\left(\dot{\overline{\theta}},\overline{\theta}\right)$] are introduced in the following manner:

$$\hat{M}(\overline{\theta}) = \begin{bmatrix} m_{11} & m_{12} \\ m_{12} & m_{22} \end{bmatrix} \tag{10}$$

$$\hat{H}\left(\dot{\overline{\theta}},\overline{\theta}\right) = \begin{bmatrix} h_{11} & h_{12} \\ h_{21} & h_{22} \end{bmatrix} \tag{11}$$

where m_1, m_2, I_{c1}, and I_{c2} correspond to the masses of the first and second links of the manipulator, and the moments of inertia of these links with respect to their mass centers indicated by C_1 and C_2, respectively. Moreover, b_1 and b_2 are used to show the viscous friction coefficients at the first and second joints. With the definitions of $d_1 = |OC_1|$ and $d_2 = |AC_2|$ as additional length parameters, the following symbols are used in Eqs. (10) and (11).

$$\begin{aligned} m_{11} &= m_1 d_1^2 + m_2\left[a_1^2 + d_2^2 + 2a_1 d_2 \cos(\theta_2)\right] + I_{c1} + I_{c2}, m_{12} \\ &= m_2 d_2\left[d_2 + a_1 \cos(\theta_2)\right] + I_{c2}, m_{22} = m_2 d_2^2 + I_{c2}, h_{11} \\ &= b_1 - 2m_2 a_1 d_2 \dot{\theta}_2 \sin(\theta_2), h_{12} = b_2 - m_2 a_1 d_2 \dot{\theta}_2 \sin(\theta_2), h_{21} \\ &= m_2 a_1 d_2 \dot{\theta}_1 \sin(\theta_2), \text{and } h_{22} = b_2. \end{aligned}$$

Substituting Eqs. (10) and (11) into Eq. (9), the following expressions are determined for T_1 and T_2, respectively:

$$\begin{aligned} T_1 = &\left[m_1 d_1^2 + m_2\left(a_1^2 + d_2^2 + 2a_1 d_2 \cos(\theta_2)\right) + I_{c1} + I_{c2}\right]\ddot{\theta}_1 \\ &+\left[m_2 d_2\left(d_2 + a_1 \cos(\theta_2)\right) + I_{c2}\right]\ddot{\theta}_2 + b_1\dot{\theta}_1 + b_2\dot{\theta}_2 \\ &-m_2 a_1 d_2\left(2\dot{\theta}_1\dot{\theta}_2 + \dot{\theta}_2^2\right)\sin(\theta_2) \end{aligned} \tag{12}$$

$$\begin{aligned} T_2 = &\left[m_2 d_2\left(d_2 + a_1 \cos(\theta_2)\right) + I_{c2}\right]\ddot{\theta}_1 + \left(m_2 d_2^2 + I_{c2}\right)\ddot{\theta}_2 \\ &+b_2\dot{\theta}_2 + m_2 a_1 d_2 \dot{\theta}_1^2 \sin(\theta_2) \end{aligned} \tag{13}$$

5. Robot manipulator control system

In order to keep the synchronization between the robot manipulator and moving belt during their engagement, it is more viable to make the control of the manipulator by considering its speed. That is, the components of the linear velocity vector of point P on the end effector of the manipulator in the horizontal plane become the parameters which should actually be controlled in a manner compatible with the command signals of the LHG law that are in the form of speed variables. On the other hand, it is easier and more practical to measure the joint speeds ($\dot{\theta}_1$ and $\dot{\theta}_2$) than the linear velocity components of point P. For this reason, an indirect control scheme is designed in the present study such that the joint speeds are selected as control variables. In this situation, it is required to express the linear velocity components of point P in terms of the joint speeds. Here, the linear position and velocity components of point P can be calculated from Eqs. (2) through (4) using the measured joint angles and their rates.

Introducing $\dot{\theta}_{1d}$ and $\dot{\theta}_{2d}$ to demonstrate the desired, or reference, joint speeds with the column matrix $\dot{\overline{\theta}}_d = \left[\dot{\theta}_{1d} \quad \dot{\theta}_{2d}\right]^T$, the error column matrix between the desired and actual joint speeds ($\overline{e}$) can be introduced as follows:

$$\overline{e} = \dot{\overline{\theta}}_d - \dot{\overline{\theta}} \tag{14}$$

In order to make the steady state errors zero, the following control law including an integral action is designating upon the torque input of the manipulator as per the computed torque method [27–29]:

$$\overline{T} = \hat{M}\ddot{\overline{\theta}}_d + \hat{H}\dot{\overline{\theta}} + \hat{K}_p\overline{e} + \hat{K}_i\int \overline{e}\,dt \tag{15}$$

where $\hat{M} = \hat{M}(\overline{\theta})$ and $\hat{H} = \hat{H}\left(\dot{\overline{\theta}}, \overline{\theta}\right)$ are defined. Also, $\hat{K}_p$ and $\hat{K}_i$ stand for the proportional and integral gain matrices, respectively, and $\hat{M}$ and $\hat{H}$ matrices are assumed to be accurately calculated. As can be seen, the proposed control system is based on the PI (proportional plus integral) control law [14, 30].

Inserting Eq. (15) into Eq. (9) and making the arrangements regarding Eq. (14), the error dynamics of the control system is obtained in the following manner:

$$\ddot{\overline{e}} + \hat{M}^{-1}\left(\dot{\hat{M}} + \hat{K}_p\right)\dot{\overline{e}} + \hat{M}^{-1}\hat{K}_i\overline{e} = \overline{0} \tag{16}$$

where $\dot{\hat{M}} = -m_2 a_1 d_2 \dot{\theta}_2 \sin(\theta_2)\begin{bmatrix} 2 & 1 \\ 1 & 0 \end{bmatrix}$.

For a finite solution, the existence of $\hat{M}^{-1}$ must be guaranteed by the invertible matrix $\hat{M}$. That is, the determinant of $\hat{M}$ must be nonzero. Here, the determinant of $\hat{M}$, i.e. $|\hat{M}|$, is obtained from Eq. (10) as follows:

$$|\hat{M}| = I_{c1}\left(m_1 d_1^2 + m_2 d_2^2 + I_{c2}\right) + m_2 a_1^2\left[m_2 d_2^2 \sin^2(\theta_2) + I_{c2}\right] + m_1 m_2 d_1^2 d_2^2 \tag{17}$$

As noticed, $|\hat{M}|$ never becomes zero unless any of the mass and inertia parameters of the links of the manipulators disappears. Because this is not possible in physical sense, $\hat{M}$ is invertible in all conditions and hence $\hat{M}^{-1}$ exists.

For a second-order two-degree-of-freedom ideal system, the error dynamics can be defined using the forthcoming expression as ω_{ci} and ζ_{ci} correspond to the bandwidth and damping parameters of the ith link (i = 1 and 2), respectively [27]:

$$\ddot{\overline{e}} + \hat{D}\dot{\overline{e}} + \hat{W}\overline{e} = \overline{0} \tag{18}$$

where $\hat{D} = \begin{bmatrix} 2\zeta_{c1}\omega_{c1} & 0 \\ 0 & 2\zeta_{c2}\omega_{c2} \end{bmatrix}$ and $\hat{W} = \begin{bmatrix} \omega_{c1}^2 & 0 \\ 0 & \omega_{c2}^2 \end{bmatrix}$.

Finally, equating Eqs. (16) and (18) to each other, $\hat{K}_p$ and $\hat{K}_i$ appear as follows:

$$\hat{K}_p = \hat{M}\hat{D} - \dot{\hat{M}} \quad (19)$$

$$\hat{K}_i = \hat{M}\hat{W} \quad (20)$$

In order to maintain the stability of the manipulator control systems throughout the engagement, the components of the matrices $\hat{K}_p$ and $\hat{K}_i$ which may be diagonal or off-diagonal are updated at certain instants.

6. Moving belt kinematics

The horizontal and lateral position components of point S (x_S and y_S) showing the slot on the moving as shown in **Figure 3** are described at the changing points symbolized by S_i (i = 1, 2, 3, and 4) by taking v_S to be constant in the next manner:

$$(x_S, y_S) = \begin{cases} x_0, y_0 + v_s(t - t_0) & , \quad x_0 \le x_S < x_1 \\ x_1 + \rho[\cos(\psi) + 1], y_1 + \rho \sin(\psi) & , \quad x_1 \le x_S < x_2 \\ x_2, y_2 - v_s(t - t_2) & , \quad x_2 \le x_S < x_3 \\ x_3 + \rho[\cos(\psi) - 1], y_3 + \rho \sin(\psi) & , \quad x_3 \le x_S < x_0 \end{cases} \quad (21)$$

In Eq. (21), the position variables of point S at the S_i location quantities are found by considering the following terms with the corresponding rotation angle at these points (ψ_i) (π = 3.14):

$$x_0 = d, y_0 = \rho - (L/2), \psi_0 = \pi \text{ rad}; x_1 = d, y_1 = (L/2) - \rho, \psi_1 = \pi \text{ rad}; x_2 = d + 2\rho, y_2 = (L/2) - \rho, \psi_2 = 0; x_3 = d + 2\rho, y_3 = \rho - (L/2), \text{and } \psi_3 = 0.$$

where x_i, y_i, and t_i denote the horizontal and lateral position variables of point S at the S_i location, and time parameter, respectively.

In Eq. (21),ψ can be determined from the following expression as a function of time (t) for $t_1 = t_0 + (|y_1 - y_0|/v_S)$, $t_2 = t_1 + (\pi\rho/v_S)$, and $t_3 = t_2 + (|y_2 - y_3|/v_S)$ with a specified t0 value:

$$\psi = \begin{cases} \pi & , \quad x_0 \le x_s < x_1 \\ \pi - [v_S(t - t_1)/\rho] & , \quad x_1 \le x_s < x_2 \\ 0 & , \quad x_2 \le x_s < x_3 \\ 2\pi - [v_S(t - t_3)/\rho] & , \quad x_3 \le x_s < x_0 \end{cases} \text{(rad)} \quad (22)$$

7. Engagement geometry

The engagement geometry between point P on the end effector of the robot manipulator and point S carried by the moving belt can be schematized on the horizontal plane as given in **Figure 4**.

Introducing v_P, γ_m, $r_{S/P}$, γ_b, and λ as the magnitude of the resulting velocity vector of point P, orientation angle of v_P from the horizontal axis, relative position of point S with respect to point P, orientation angle of v_S from the horizontal axis, and angle between $r_{S/P}$ and horizontal axis in **Figure 4**, respectively, v_P, γ_b, and λ can be calculated using the equations below:

$$v_P = \sqrt{\dot{x}_P^2 + \dot{y}_P^2} \quad (23)$$

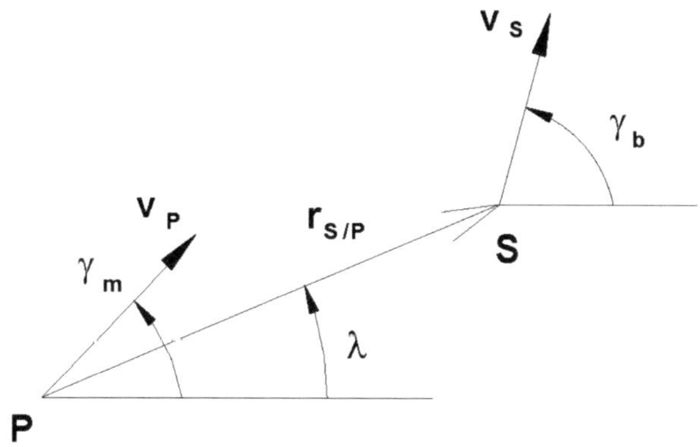

Figure 4.
Engagement geometry between the tip point of the manipulator and slot on the belt.

$$\gamma_b = \begin{cases} \pi/2 & , \quad x_0 \le x_S < x_1 \\ \psi - (\pi/2) & , \quad x_1 \le x_S < x_2 \\ -\pi/2 & , \quad x_2 \le x_S < x_3 \\ \psi - (\pi/2) & , \quad x_3 \le x_S < x_0 \end{cases} \text{(rad)} \tag{24}$$

$$\lambda = a\tan\left[(y_S - y_P)/(x_S - x_P)\right] \tag{25}$$

8. Guidance law

In the LHG law, it is intended to keep the end effector of the manipulator always on the collision triangle that is formed by the end effector, slot, and predicted intercept point. For this purpose, the most convenient approach is to orient the velocity vector of point P on the end effector ($\vec{v}_{Pactual}$) towards the predicted intercept point (I) at which the collision between the end effector and slot is going to be happen after a while as depicted in **Figure 5** with $\vec{v}_S$ and $\vec{v}_{Pideal}$ which denote the velocity of point S and the ideal velocity of point P [26].

In this law, in order for point P to catch point S, the guidance command (γ_m^c) is derived as follows [4, 26]:

$$\gamma_m^c = \lambda + a\sin\left[(v_S/v_P)\sin(\gamma_b - \lambda)\right] \tag{26}$$

Here, using the measurements of v_S by means of the appropriate sensors on the belt rollers, the position variables x_S, y_S, and ψ are obtained.

In the application, the following column matrix including the reference values of the linear velocity components of point P ($\dot{\bar{r}}_{Pd}$) are formed using γ_m^c:

$$\dot{\bar{r}}_{Pd} = v_P\left[\cos(\gamma_m^c) \quad \sin(\gamma_m^c)\right]^T \tag{27}$$

In order to overcome the algebraic loop which occurs because the values of v_P and $\dot{\bar{r}}_d$ are dependent on each other, a nonzero value which is compatible with the current component-picking motion of the manipulator is assigned to v_P at the initiation of the engagement.

The guidance commands can be expressed in terms of the angular speeds by means of the next expression regarding Eqs. (6) and (27):

$$\dot{\bar{\theta}}_d = \hat{J}_P^{-1}\dot{\bar{r}}_{Pd} \tag{28}$$

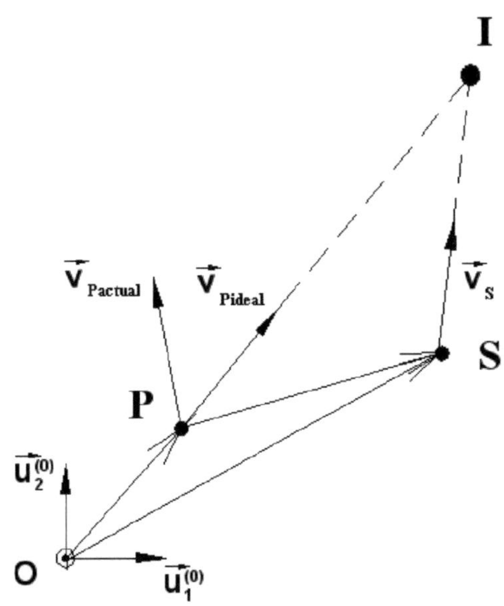

Figure 5.
Linear homing guidance law geometry.

9. Computer simulations

The numerical values considered in the computer simulations are submitted in **Table 1** along with the engagement block diagram in **Figure 6**.

The unit step responses at the first and second joints of the manipulator are submitted in **Figures 7** and **8** in which the discrete and continuous lines show the desired, or reference, and actual values of the joint angles, respectively. As shown, the desired joint speeds can be caught within the assigned bandwidth.

Parameter	Value	Parameter	Value	Parameter	Value
a_1 and a_2	1.25 m	b_1 and b_2	0.001 N·m·s/rad	L	2 m
d_1 and d_2	0.625 m	ω_{c1} and ω_{c2}	62.832 rad/s (=10 Hz)	ρ	0.5 m
m_1 and m_2	10 kg	ζ_{c1} and ζ_{c2}	0.707	d	1.5 m
I_{c1} and I_{c2}	1.302 kg·m^2				

Table 1.
Numerical values used in the computer simulations.

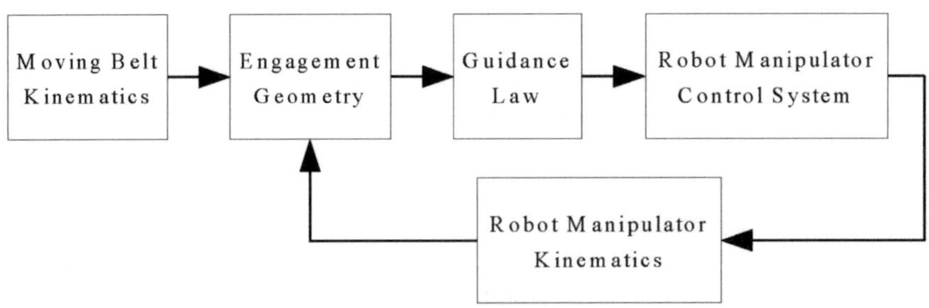

Figure 6.
Block diagram for the robot manipulator-moving belt.

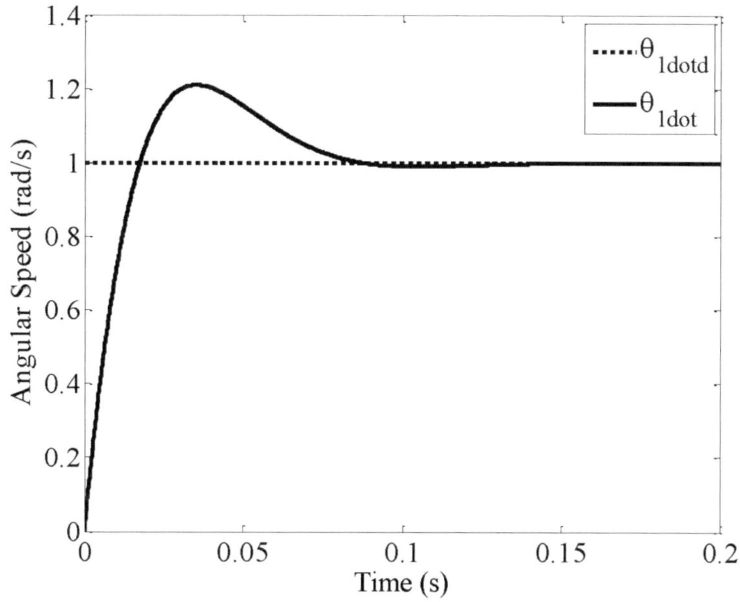

Figure 7.
Unit step response of the control system at the first joint.

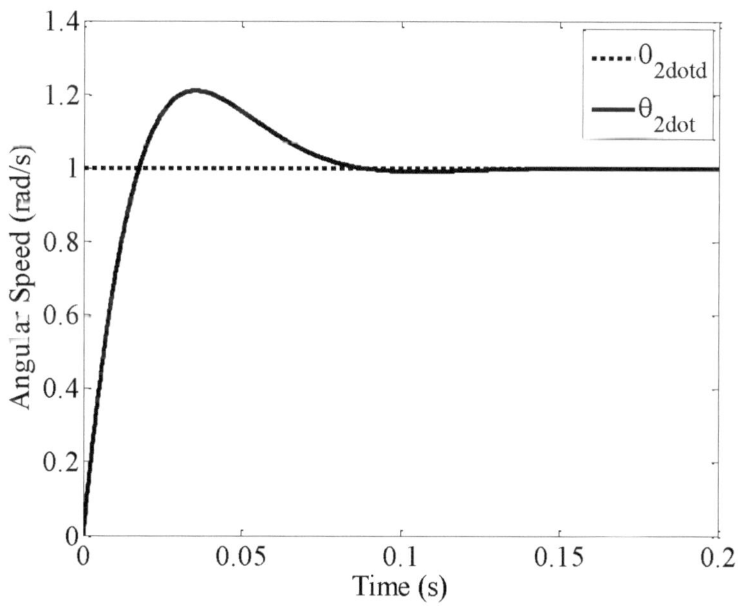

Figure 8.
Unit step response of the control system at the second joint.

In the designated engagement scenarios, it is assumed that the slot on the moving belt stands at point S_O with x_{SO} = 1.5 and y_{SO} = −0.5 m at the instant when the robot manipulator is at rest. Furthermore, ramp-type angular speed inputs are applied to the manipulator joints in order for point P to attain its initial engagement velocity (v_{Pe}) at the end of 0.1 s. Here, the maximum angular speeds of the direct current (DC) electric motors connected to the joints are taken as 20 rad/s. The disturbance due to the nonlinear friction and noise on the sensors on the joints are

Conf. No.	Robot manipulator			Moving belt velocity (v_S) (m/s)
	Initial position of the tip point (m)		Velocity at the beginning of the engagement (v_{Pe}) (m/s)	
	x_{PO}	y_{PO}		
1	−0.5	−0.5	5.0×10^{-3}	0.5
2	−0.5	−0.5	5.0×10^{-3}	1.0
3	−0.5	−0.5	0.5	1.0
4	−1.0	−0.5	0.5	1.0
5	−1.0	−0.5	5×10^{-5}	2.5

Table 2.
Simulation configurations considered.

randomly varied in the ranges of ± 10 N·m and $\pm 1 \times 10^{-3}$ rad, respectively. The solver is selected to be the ODE5 (Dormand-Prince)-type solver with a fixed time step of 1×10^{-4} s. The simulation configurations are designated as in **Table 2** along with the numerical values of the related parameters.

Having performed the computer simulations performed in the MATLAB® SIMULINK® environment, the results given in **Table 3** are attained. As samples, the engagement geometry for the configuration number 1 is given in **Figure 9** along with the plots for the changes of the velocity of point *P*, joint angles, joint speeds, and joint accelerations are submitted in **Figures 10–13**, respectively. Moreover, the engagement geometries for the sample configurations are plotted in **Figures 14** and **15**.

10. Discussion

As given in **Figures 9, 14**, and **15** which belong to the designated simulation configurations at belt speeds from 0.5 to 2.5 m/s, it is observed that the tip point of the manipulator can catch the slot on the moving belt even at higher speeds. In the present work, the slot is caught by the manipulator near the left side of the belt. Actually, this placement strategy is desired in order to diminish the power consumption of the robot manipulator by keeping the motion distance short compared to the distance to the right side of the belt. Looking at the simulation data which are presented in the forms of relevant kinematic parameters of the manipulator in **Figures 10–13**, it can be verified that the angular speed values required at the joints of the manipulator can be attained even with industrial DC electric motors as well as the angular excursion demands.

11. Conclusion

Motion planning constitutes one of the significant issues in the development of autonomous system. In this context, guidance concept has been applied on munition developed to satisfy precise hitting requirements for recent years. Both theoretical studies and field implementations have revealed that the guidance algorithms have led the relevant munition to the desired target points successfully. Of course, the performance of the designated guidance scheme is directly related to the control systems whose primary function is to obey the commands generated by the

Conf. No.	Engagement time (s)	Maximum tip Point velocity (v_{Pmax}) (m/s)	Joint angles (°)				Joint speeds (rad/s)			
			θ_1		θ_2		$\dot{\theta}_1$		$\dot{\theta}_2$	
			Min.	Max.	Min.	Max.	Min.	Max.	Min.	Max.
1	0.299	31.339	−61.769	44.093	−159.876	−104.421	−6.127	20.161	−6.651	20.085
2	0.308	30.082	−62.451	52.147	−162.918	−106.102	−6.573	20.288	−8.469	20.202
3	0.267	30.654	−61.429	50.059	−161.766	−105.679	−6.268	21.158	−8.972	20.602
4	0.327	29.050	−90.000	54.134	−164.874	−106.239	−6.979	20.189	−17.313	20.106
5	0.745	25.000	−90.000	180.000	−164.369	−46.956	−20.146	0.000	−9.915	20.091

Table 3.
Results attained from the considered simulation configurations.

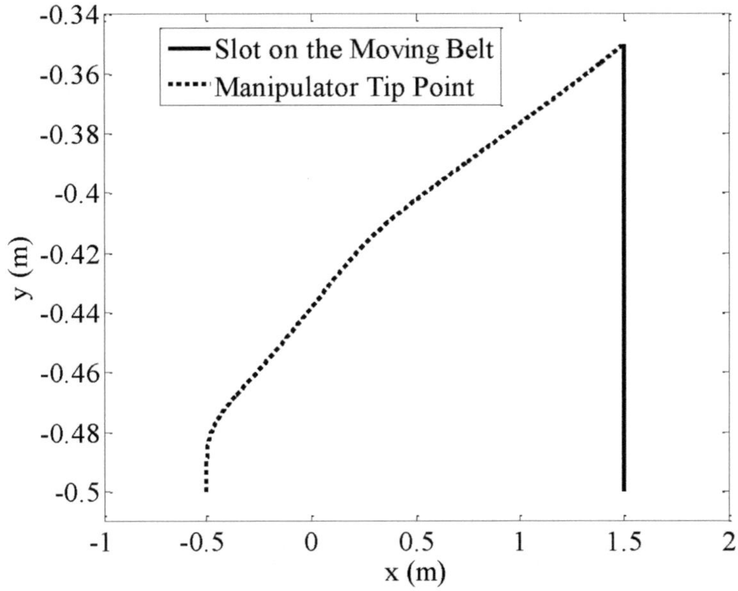

Figure 9.
Engagement geometry for the initial position components of the tip point of $x_{Po} = -0.5$ *m and* $y_{Po} = -0.5$ *m with a moving belt velocity of 0.5 m/s.*

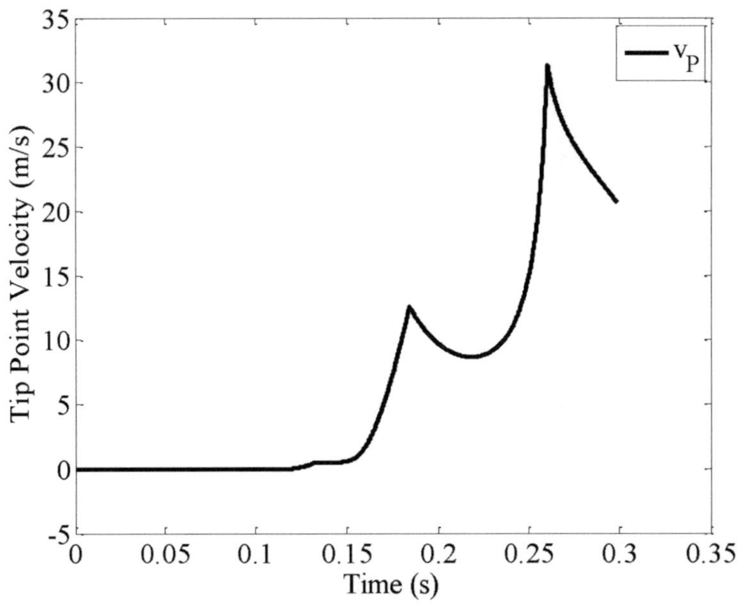

Figure 10.
Change of the velocity of point P in time for the initial position components of the tip point of $x_{Po} = -0.5$ *m and* $y_{Po} = -0.5$ *m with a moving belt velocity of 0.5 m/s.*

guidance law. For this purpose, several guidance and control approaches are proposed depending on the kind of the planned mission as can be encountered in the related literature.

Regarding the fact that robot manipulators are designed to achieve certain tasks which are usually specified before the execution, it can be a viable way to apply the

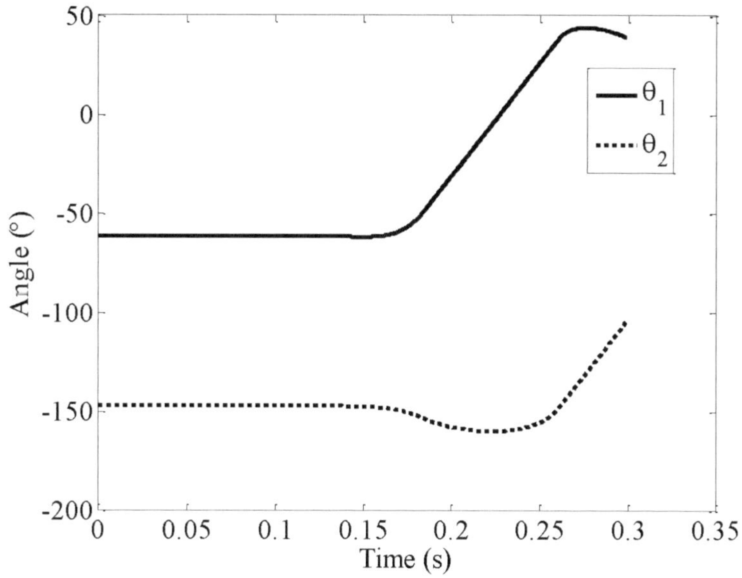

Figure 11.
Change of the joint angles in time for the initial position components of the tip point of $x_{Po} = -0.5$ *m and* $y_{Po} = -0.5$ *m with a moving belt velocity of 0.5 m/s.*

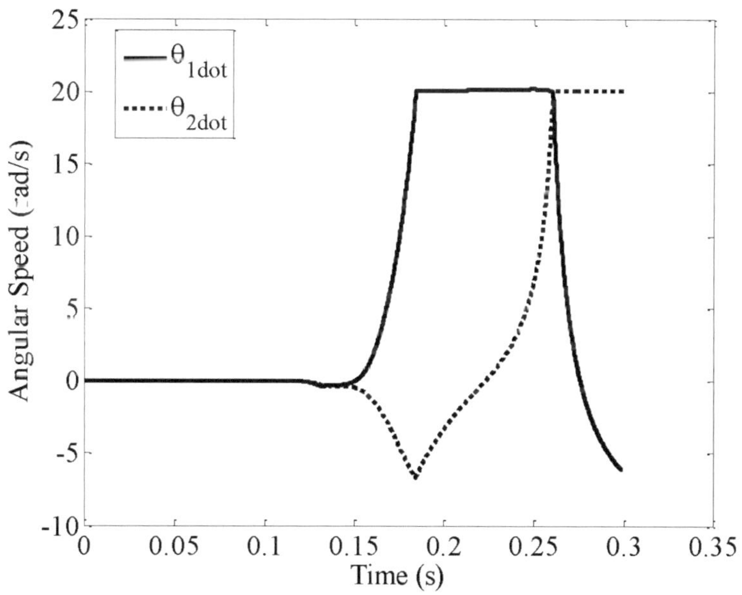

Figure 12.
Change of the joint speeds in time for the initial position components of the tip point of $x_{Po} = -0.5$ *m and* $y_{Po} = -0.5$ *m with a moving belt velocity of 0.5 m/s.*

similar approach in munition for the motion planning tasks of the manipulators. As a result of the present work, it can be deduced that the guidance-based approach leads to a successful placement for the components onto the intended slots in continuous engagement operations. This can be done even under considerable disturbing effects and undesirable changing speed conditions of the belt with lower power consumption levels.

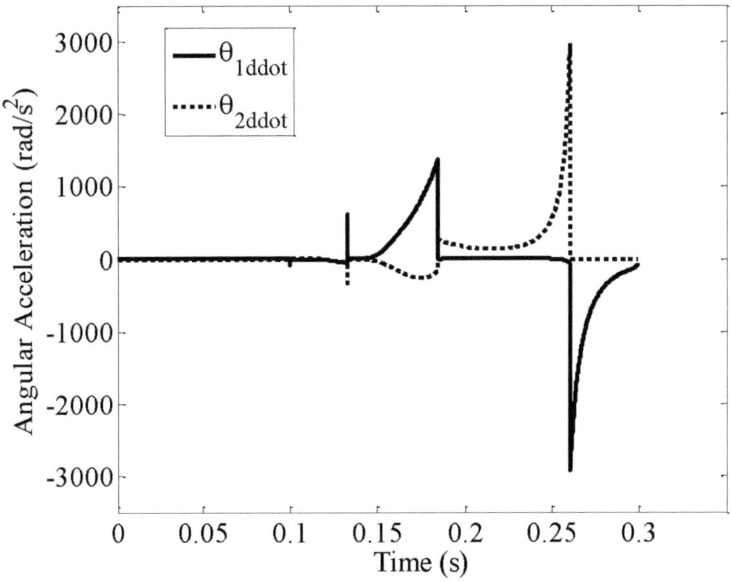

Figure 13.
Change of the joint accelerations in time for the initial position components of the tip point of $x_{Po} = -0.5$ *m and* $y_{Po} = -0.5$ *m with a moving belt velocity of 0.5 m/s.*

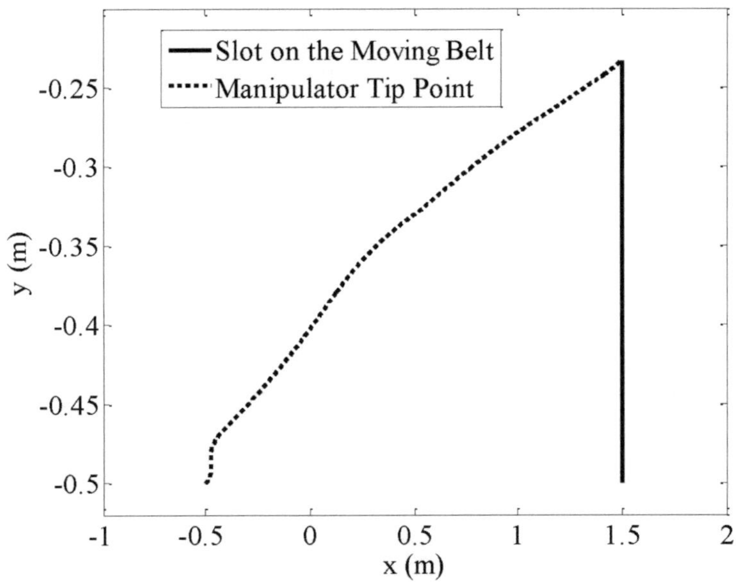

Figure 14.
Engagement geometry for the initial position components of the tip point of $x_{Po} = -0.5$ *m and* $y_{Po} = -0.5$ *m with a moving belt velocity of 1.0 m/s.*

Although the applicability of the guidance and control approach on the robot manipulators is demonstrated by means of relevant computer simulations, there is not seen any serious difficulty to adapt the suggested concept into practice. This way, some of the robotic operations can be performed in an efficient manner.

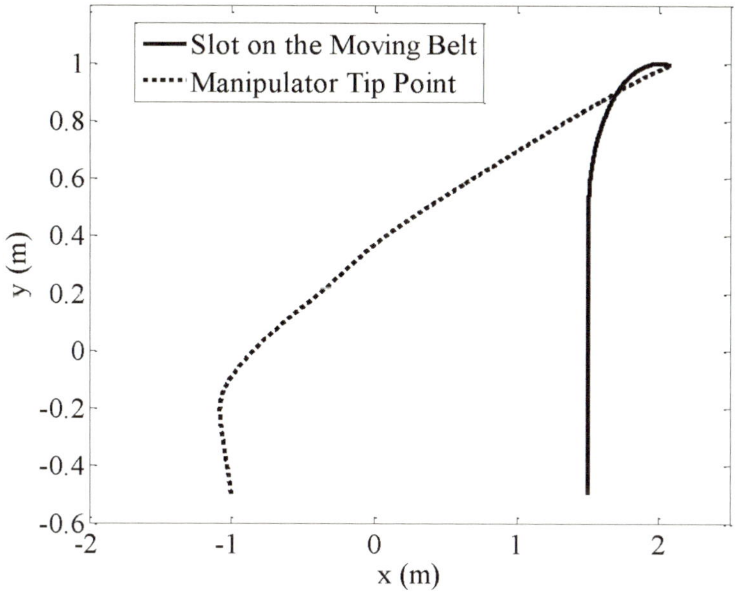

Figure 15.
Engagement geometry for the initial position components of the tip point of $x_{P_O} = -1.0$ *m and* $y_{P_O} = -0.5$ *m with a moving belt velocity of 2.5 m/s.*

Author details

Bülent Özkan
Mechanical Engineering Department, Gazi University, Ankara, Turkey

*Address all correspondence to: bozkan37@gmail.com

References

[1] Lan TS, Yeh LJ, Chiu MC, Hwang YX. Mint: Construction of the control system of cleaning robots with vision guidance. Mathematical Problems in Engineering. 2013;**2013**:1-6

[2] Ni T, Zhang H, Xu P, Yamada H. Mint: Vision-based virtual force guidance for tele-robotic system. Computers and Electrical Engineering. 2013;**39**:2135-2144

[3] Lewis FL. Abdallah. Dawson, DM. Control of Robot Manipulators: CT; 1993

[4] Özkan B. Guidance and control of a planar robot manipulator used in a mounting line. In: Proceedings of the 18th IFAC World Congress; 2011; Milan. Italy.

[5] Spong MW. Motion Control of Robot Manipulators-Control Handbook The Coordinated Science Laboratory. USA: University of Illinois at Urbana-Champaign; 2003

[6] İslamoğlu NE, Ryu K, Moon I. Mint: Labour productivity in modular assembly: a study of automotive module suppliers. International Journal of Production Research. 2014;**52**(23): 6954-6970

[7] Jang JS, Rim SC, Park SC. Mint: Reforming a conventional vehicle assembly plant for job enrichment. International Journal of Production Research. 2006;**44**(4):703-713

[8] Mehrandezh, M, Sela, MN, Fenton, RG, Benhabib, B. Proportional navigation guidance in robot trajectory planning for intercepting moving objects. In: Proceedings of the 1999 IEEE International Conference on Robotics and Automation; Detroit, Michigan, USA; 1999. p. 145–150

[9] Zhang Z, Wang WQ, Siddiqui S. Predictive function control of a two-arm robot manipulator. In: Proceedings of the IEEE International Conference on Mechatronics and Automation 2004-2009; 2005; Niagara Falls. Canada.

[10] Ferrara, A, Scattolini, R. Control of a robot manipulator for aerospace applications. In: Proceedings of the 6th Dynamics and Control of Systems and Structures in Space (DCSSS); Riomaggiore, Italy; 2004.

[11] Kunwar F, Benhabib B. Mint: Advanced predictive guidance navigation for mobile robots: a novel strategy for rendezvous in dynamic settings. International Journal on Smart Sensing and Intelligent Systems. 2008; **1**(4):858-890

[12] Mehrandezh, M. Navigation-guidance-based robot trajectory planning for ınterception of moving objects [thesis]. PhD Thesis, University of Toronto, Canada; 1999.

[13] Mehrandezh, M, Sela, NM, Fenton, RG, Benhabib, B. Mint: Robotic interception of moving objects using an augmented ideal proportional navigation guidance technique. IEEE Transcations on Systems, Man, and Cybernetics-Part A: Systems and Humans. 2000:30:3:238–250.

[14] Caner U, Eroğlu M. Mint: State feedback plus ıntegral error controller approach for robot arm control design (in Turkish). Journal of the Faculty of Engineering and Architecture of Gazi University. 2004;**19**(3):335-342

[15] Zarchan, P. Tactical and Strategic Missile Guidance. Vol. 157. Progress in Aeronautics and Astronautics, AIAA. Washington DC; USA; 1994.

[16] Koren Y, Shoham M. Mint: End-effector guidance of robot arms. Annals of the CIRP. 1987;**36**(1):289-292

[17] Garcia, GJ, Gil, P, Llácer, D, Torres, F. Guidance of robot arms using depth data from RGB-D camera. In: ICINCO 2013-10th International Conference on Informatics in Control, Automation and Robotics; 2013; Reykjavík, Iceland; p. 315–321.

[18] Becker BC, Voros S, MacLachlan RA, Hager GD, Riviere CN. Active guidance of a handheld micromanipulator using visual servoing. In: Proceedings of the IEEE International Conference on Robotics and Automation-2009 (ICRA'09); 2009; Kobe. Japan.

[19] Park JG, Shin JH. Mint: Autonomous navigation for a mobile robot using navigation guidance direction and fuzzy control. The Transactions of the Korean Institute of Electrical Engineers. 2014; **63**(1):108-114

[20] Belkhouche F, Rastgoufard P, Belkhouche B. Mint: Robot navigation-tracking of moving objects using the standard proportional navigation law. IEEE Trans. Robotics. 2007;**01**

[21] Seeraji S, Ovy EG, Alam T, Zamee A, Emon ARA. Mint: A flexible closed loop PMDC motor speed control system for precise positioning. International Journal of Robotics and Automation. 2011;**2**(3):211-219

[22] Azimi V, Menhaj MB, Fakharian A. Mint: Tool position tracking control of a nonlinear uncertain flexible robot manipulator by using robust H2/H∞ controller via T-S fuzzy model. Sadhana. 2015;**40**(2):307-333

[23] Azimi, V, Simon, D, Richter, H. Stable robust adaptive impedance control of a prosthetic leg. In: Proceedings of the ASME Dynamic Systems and Control Conference. Columbus. USA: Ohio; 2015

[24] Slotine JJE, Coetsee JA. Mint: Composite adaptive controller of robot manipulators. Automatica. 1989;**25**(4): 509-519

[25] Slotine JJE, Li W. Mint: Adaptive manipulator control: a study case. IEEE Transactions on Automatic Control. 1988;**33**(11):995-1003

[26] Özkan B, Özgören MK, Mahmutyazıcıoğlu G. Implementation of linear homing guidance law on a two-part homing missile. In: Proceedings of the 17th IFAC World Congress; 2008; Seoul. Republic of Korea.

[27] Ogata, K. Modern Control Engineering. 2nd ed.; Prentice-Hall International Editions; 1990.

[28] Bevly D, Dubovsky S, Mavroidis C. Mint: A simplified cartesian-computed torque controller for highly geared systems and its application to an experimental climbing robot. Journal of Dynamic Systems, Measurement, Control. 2000;**122**

[29] Yang Z, Wu J, Mei J, Gao J, Huang T. Mint: Mechatronic model based computed torque control of a parallel manipulator. International Journal of Advanced Robotic Systems. 2008;**5**(1):123-128

[30] Koca H, Doğan M, Taplamacıoğlu MC. Mint: Cartesian-specific control of a robot manipulator (in Turkish). Journal of the Faculty of Engineering and Architecture of Gazi University. 2008;**23**(4):769-776

Chapter 6

3D Printed Walking Robot Based on a Minimalist Approach

Ivan Chavdarov

Abstract

3D printing technology enables the design and testing of highly complex robot prototypes and joints. Here an original idea for a walking robot is presented, based on a minimalist approach. Although the robot has a simple mechanical structure using only 2 motors, it can walk, turn around its central axis and climb high obstacles. The simple design ensures higher reliability in terms of mechanics and control. A design principle is suggested, which minimizes power consumption during climbing. The kinematics and static conditions for overcoming an obstacle are analyzed and the movements of the robot are simulated. A 3D-printed prototype of the robot is created. It is used for experiments to test the efficiency of different materials and shapes for the robot's feet when climbing. The results are ranked and compared with the efficiency of other walking robots.

Keywords: Walking robot, Robot design, Overcoming an obstacle, 3D print, Minimalist approach

1. Introduction

Walking robots are designed to move in an environment with multiple and diverse obstacles [1]. For that reason they need to do complex coordinated motions, which require a complex mechanical structure and advanced control system with multiple sensors [2]. Mobile robots, created to conduct rescue operations or inspection tasks in an urban environment, often face problems when they need to climb stairs [3]. As opposed to wheel robots, walking robots have a more complex design, more motors and are slower [1, 3]. Often, they have more degrees of freedom and use special algorithms. The advantage is that the robot can do complex movements [1, 4]. However, this comes at the price of more components in the design. Hence, the disadvantages:

- the complex structure is expensive
- maintenance is difficult
- it is more difficult to be controlled
- the probability that a fault occurs increases with the number of components
- the higher number of motors means more energy is consumed and the total mass of the robot increases.

The following questions arise: What would be the simplest walking robot design that can effectively overcome obstacles? What would be the minimal number of degrees of freedom for such a robot? Can a simple control system work when overcoming different types of obstacles? How can 3D printing technology bring additional advantages in the development of robots, based on minimalist approach?

The stability of a walking robot is a major issue, because it defines the conditions under which it will not lose balance. There are two types of stability - static and dynamic. Static walking means that the robot can be stopped at any moment during the gait cycle without losing balance. Dynamic walking means that additional internal movements and algorithms are needed to sustain balance.

Two-legged robots usually have dynamic stability and a relatively large number of degrees of freedom [5–7]. They can go around or climb obstacles, but need a complex control system and consume a lot of energy. Their reliability is lower, due to the large number of electrical and mechanical components. There are experimental two-legged robots which can sustain static balance.

Alternative design solutions with a minimum number of mechanical elements [8] and nature-inspired robots are being sought [9]. In [10] is presented an ultra-light, inexpensive two-legged robot "SLIDER" with a design of the leg without a knee. This non-anthropomorphic design with straight legs reduces the weight of the legs significantly, while maintaining the same functionality as anthropomorphic legs. The robot has 8 degrees of freedom, four for each leg.

The four-legged 3D printed robot presented in [9] is 3D printed with PLA (polylactic acid). It has a simple design and can walk without any form of software or controller. The robot consists of a rectangular body and four legs, each with a degree of freedom that rotates and raises the leg. At the end of each of the legs is mounted a rubber foot to improve traction. Although there are only 4 degrees of freedom, the robot realizes a gait which is similar to the gaits used by walking primates and cattle (grazing animals).

In [8] is presented a robot with one motor and several clutches. By sequential action of the clutches, the proposed robot can rotate in different directions and can walk. It can be combined with other identical modules to build more complex reconfigurable robots.

Walking mechanisms that do not need motors are studied [11, 12]. However, their passive movement is realized only on slopes and is difficult to control.

3D printing technology is used to create and test the qualities of prototypes of walking robots [9, 10, 13, 14]. Conventional materials such as PLA [9] and ABS (Acrylonitrile Butadiene Styrene) [13] are most commonly used. In [14] a methodology for 3D printing of hermetic soft drives with built-in air couplings is proposed. Two materials are used, hard and flexible, and printing is done with a printer with two extruders. Additive manufacturing is evolving and finds more and more applications in robotics. In [15] the main focus is on developing a methodology for creating a 3D printed, low-budget robotic arm with six degrees of freedom that can be used with an external artificial intelligence system. In [16] is used a custom 3D printer and CAD model of a structure for a specialized device, which consists of two-layer micro actuators driven by hydrogels.

Maintaining stability when moving [17, 18] and overcoming obstacles [2, 19] are also important issues that have been studied in recent years.

For these reasons, here it will be discussed the design of a new 3D printed model of a walking robot, based on a minimalist approach [20, 21]. Mies van der Rohe's motto "Less is more" reflects the approach to the robot's design. Using only two motors, the robot can walk forward and backward, rotate 360 deg. around itself and overcome obstacles including climbing stairs.

2. Simple mechanical design

It is well known that a robot needs at least 6 degrees of freedom to reach any point in its workspace with any orientation. 3 for changing the position and 3 for realizing the random orientation. Since the walking robot moves on a surface, it can be concluded that 3 degrees of freedom are enough - X, Y axis and orientation. After all there are examples of mobile robots with two motors that achieve satisfactory results 1. A new simple design of a two-motored robot, called "Big Foot", is suggested.

The robot's body is made up of a round base {1} and a platform {2} in which all the main elements are located. The platform is mounted in the center of the circular base, and the two bodies can rotate relative to each other around the vertical axis R1 (see **Figure 1**). The movement around R1 is realized by means of a controlled motor {6}. The stator of this motor is fixed on the platform {2}, and the rotor is connected by means of a reducer to the base {1}. The motor {7} is located in the platform {2} and drives the shaft {8} by means of a gear mechanism. This shaft performs the second important rotation R2, which is perpendicular to R1. Two arms {3} are fixed to the shaft {8}, and two feet {4} are mounted at the ends of the arms. For proper walking, the feet {4} and the round base {1} need to move with a constant orientation with respect to each other. To achieve this, a gear mechanism {5} is used, which has a gear ratio of 1. It consists of 3 gears with the same module and number of teeth which are mounted in the arm {3} (Figur 1) The 3D printed model is powered by a rechargeable battery, and the control is carried out remotely via Bluetooth communication with a PC or a smartphone. Different variants of the control software are developed using sensors of different types. Video with the robots movements is available from: Video 1.

The key elements for walking are the body {2}, arms {3} and feet {4}. This is the basic structure of the robot. While walking, the body and feet remain parallel **Figure 2**.

Initially the body {2} is fixed (**Figure 2a**). The arm {3} is rotating and thanks to the gears z_1, z_2, z_3 the feet are moving parallel to the fixed body before reaching the ground. Afterwards, the feet {4} are fixed (**Figure 2b**), the arm {3} is rotating and this time the body {2} is in motion, remaining parallel to the feet. The trajectories can be seen in the following videos: Video 2 and Video 3.

The trajectory of any edge point of gear z_2 is interesting. The trajectory resembles a heart and is called the cardioid - a type of cycloid. It can be followed in the animation: Video 4.

The rotation mechanism is presented in **Figure 3**. Here the gear motor {6} works in a mode where the rotor is fixed and the body rotates as the stator operates. Thus the robot can rotate to any angle without any of the wires tangling up.

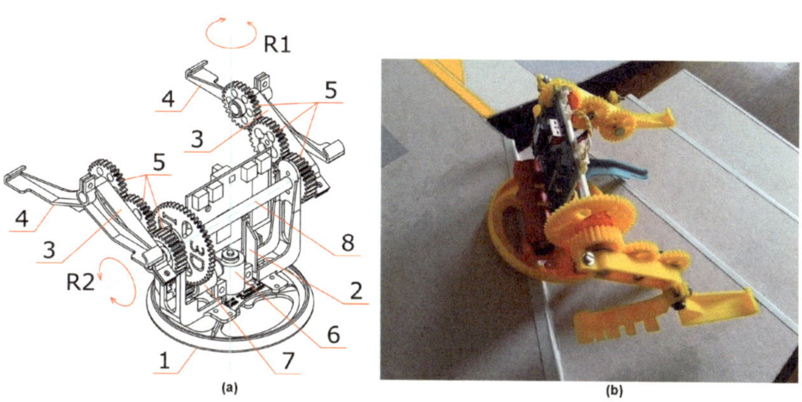

Figure 1.
Structure of the big foot robot and a picture of the 3D printed prototype.

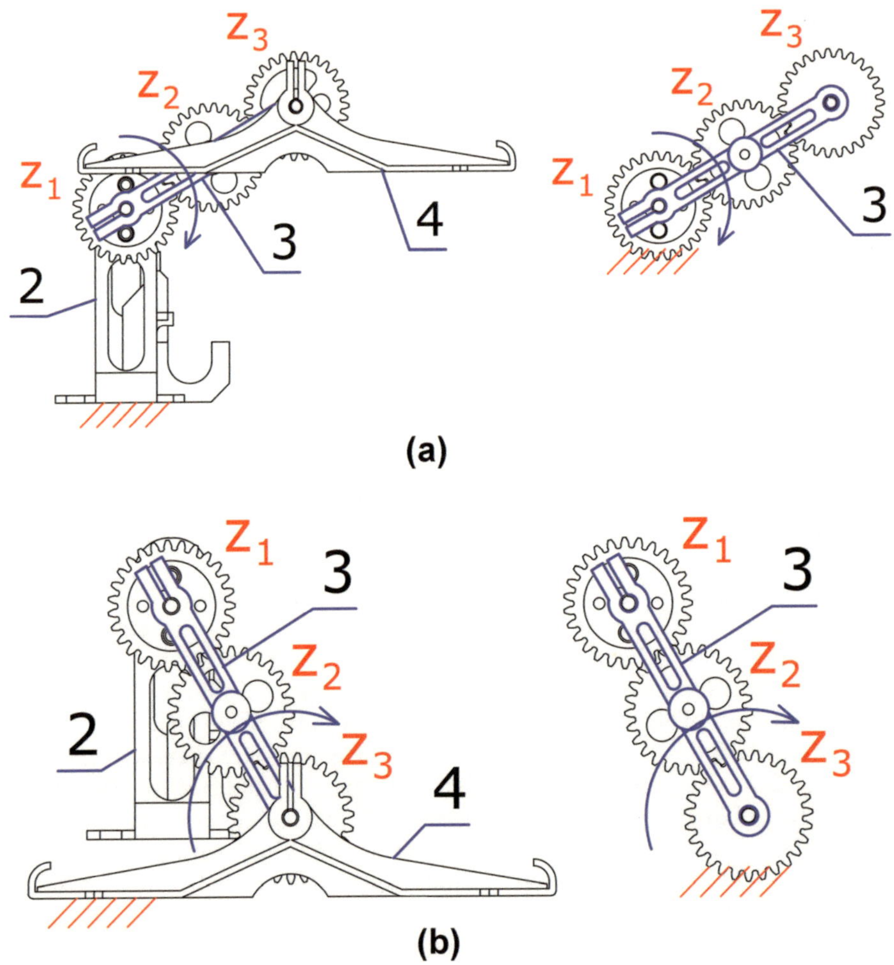

Figure 2.
The mechanism for maintaining a parallel movement between the body and the feet. a) Fixed body {2}. b) Fixed feet {4}.

Walking on flat terrain is accomplished by repeating between two phases:

Phase 1 – the two feet {4} are acting as supports. The motor {6} by means of the shaft {8} drives the arms {3}, which rotate around point A. The body of the robot is moved, where all its points move along the trajectories of arcs of a circle with radius $R_{AB} = \overline{AB} = L_3$ and angle $\varphi_B = \alpha_{max} - \alpha_{min}$. The body of the robot travels forward with one step S (**Figure 4**).

To simplify the theoretical model, it is assumed that the mass of all moving parts during this phase is concentrated at point C1. The coordinates of this point (mass center) are given in **Figure 4**. When designing the robot, it is aimed to keep the center of gravity C1 as low as possible. This increases the stability of the robot.

The horizontal movement of the robot's body is evaluated by:

$$X_{c1} = X_B = X_A + L_3 \cos(\alpha), \tag{1}$$

X_A is the horizontal coordinate of point A with respect to a fixed coordinate system, and $\alpha = \alpha(t)$ is the current angle of rotation of the unit {3} with length L_3 with respect to the horizon. The vertical displacement of point C1 is determined by:

$$Y_{c1} = Y_B - h_{c1} = Y_A + L_3 \sin(\alpha) - h_{c1}. \tag{2}$$

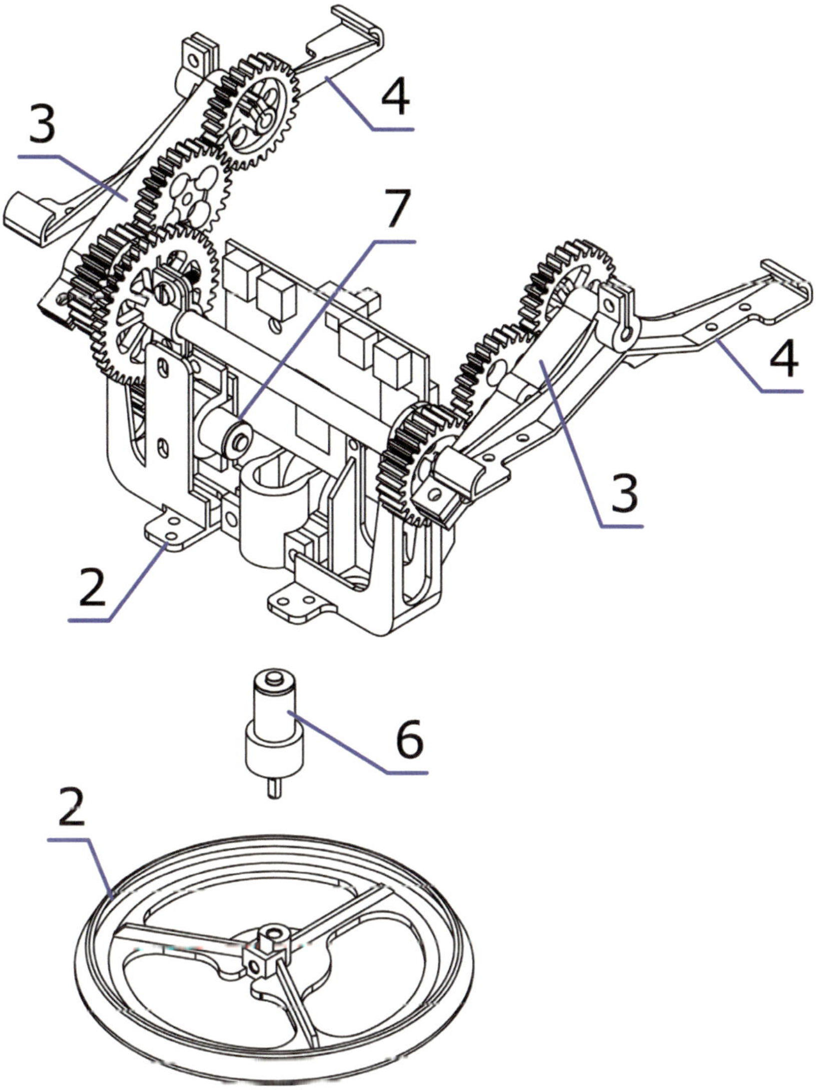

Figure 3.
Rotating mechanism.

Where Y_B is the vertical coordinate of point B, h_{c1} is the vertical distance to the center of mass at point C_1. The height h_{c1} does not change during movement. The robot moves at low speed and therefore the inertial forces are not taken into account. The torques at points A and B are determined by:

$$M_{A1} = [L_3 \cos(\alpha) + d]G_1, \tag{3}$$

$$M_{B1} = dG_1, \tag{4}$$

$G_1 = m_1 g$ is the robot's body weight, and g is the Earth's gravitational acceleration.

After differentiating (1) and (2) is obtained the velocity of the robot:

$$\left| \begin{array}{l} V_x = \dot{X}_{C1} = \dot{\alpha} L_3 \sin(\alpha) = \omega L_3 \sin(\alpha) \\ V_y = \dot{Y}_{C1} = -\dot{\alpha} L_3 \cos(\alpha) = -\omega L_3 \cos(\alpha) \end{array} \right. \tag{5}$$

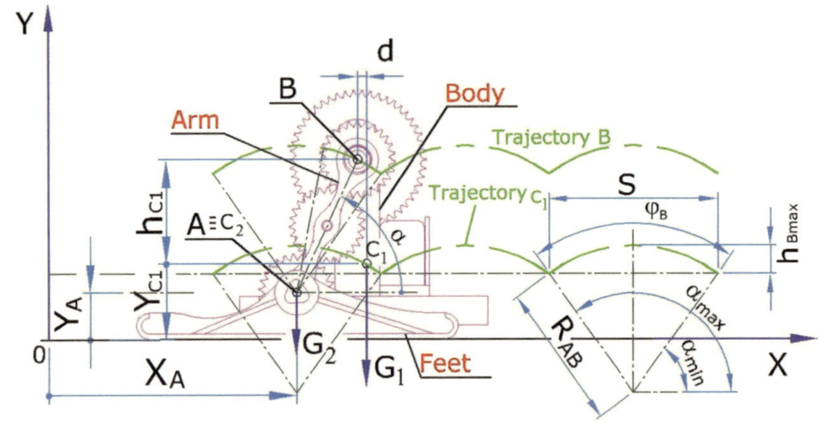

Figure 4.
Phase of support steps.

Here ω is the angular velocity of the arms {3}. From (5) the magnitude and velocity of a point from the robot's body can be determined. This phase ends when the round base {1} reaches the ground. Then the body stops moving and a second phase beggins.

Phase 2 - The robot body is stationary, the arms {3} rotate about an axis at point B, the feet {4} are moving. The mass of the moving parts is less than the mass of the robot's body. It is assumed that it is concentrated at point B. During this phase, the robot does not move and therefore its speed is zero. The feet move progressively along a trajectory, which is an arc of a circle. The torques at points A and B are determined by:

$$M_{A2} = 0, \tag{6}$$

$$M_{B2} = L_3 G_2 \cos(\alpha), \tag{7}$$

$G_2 = m_2 g$ is the mass of the moving elements in this phase. The loading in the shafts is cyclic, with shaft A being more loaded (see formulas (3), (4), (6) and (7). During the transition from phase 1 to phase 2 and vice versa, shock loads occur in the construction of the robot, which are not taken into account.

3. Determining the basic dimensions

When walking on a relatively flat ground the robot switches between two phases where the contact area with the ground is large. Movement is balanced and reliable. In **Figure 5** are presented the basic dimensions of the 3D printed prototype. Five lengths (L_1-L_5; **Figure 3**) and their proportions determine the qualities of the robot and its capability to walk and overcome obstacles.

Obviously, the larger the model, the higher the obstacles that it can overcome. Therefore, the height of the obstacle h_o should be compared with the height—H_R and the length—B_R of the robot. Thus, different designs of one robot and even a variety of different robots can be objectively compared.

A dimensionless coefficient is suggested with the help of which the scale of a robot and an obstacle can be compared:

$$K_{ro} = \frac{h_o}{\sqrt{H_R B_R}}. \tag{8}$$

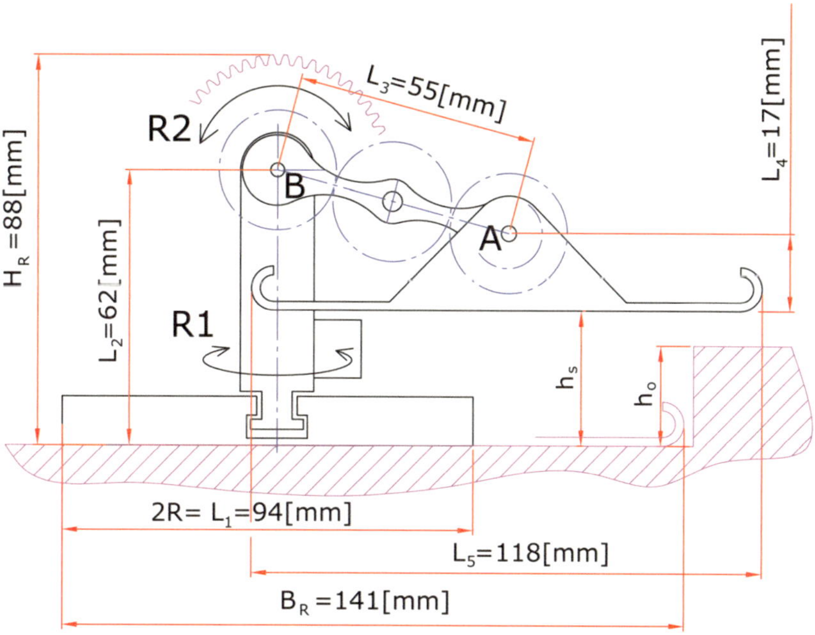

Figure 5.
Basic dimensions of the 3D printed prototype.

H_R and B_R are dimensions according to **Figure 5**. The coefficient usually assumes positive values less than one. $K_{ro} > 1$ only for climbing and jumping robots.

In order for the robot to move, it is necessary for the body and its feet to reach the ground. This is only possible under certain conditions for the sizes L_2, L_3 and L_4. The conditions are set by the inequalities:

$$\left| \begin{array}{l} L_2 \le L_3 + L_4 \\ L_4 \le L_2 + L_3 \end{array} \right. \tag{9}$$

Dimensions L_1 and L_5 are important for increasing the robot's stability, but their excessive increase reduces the maneuverability of the robot and increases its overall dimensions.

From **Figure 6** can be determined the step S, at which the robot moves

$$S = 2\sqrt{L_3^2 - (L_2 - L_4)^2} \tag{10}$$

Maximum lift height of the body h_{Bmax}

$$h_{Bmax} = L_3 - L_2 + L_4 \tag{11}$$

Maximum lift height of the feet h_{Smax}

$$h_{Smax} = L_2 + L_3 - L_4 \tag{12}$$

During the phase of support feet (phase 1) the arm is rotated at an angle φ_B:

$$\varphi_B = 2artan\left(\frac{S}{2(L_2 - L_4)}\right) \tag{13}$$

Accordingly, in phase 2, the arm {3} rotates at an angle $\varphi_s = 2\pi - \varphi_B$.

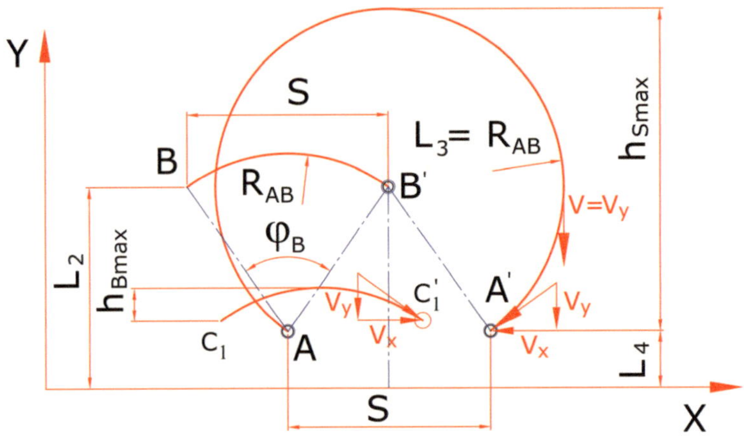

Figure 6.
Scheme for determining the geometric parameters for walking on flat terrain.

When attacking an obstacle with the robot's body, the height of the obstacle h_0 must be less than the maximum possible lifting of the robot body ($h_0 < h_{Bmax}$). When attacking with the feet {4} the maximum height of the obstacle h_0 is determined in a similar way $h_o < h_{Smax}$.

After differentiating formula (10) with respect to L_2 and knowing the values of lengths L_3 and L_4, is obtained

$$\frac{dS}{dL_2} = -\frac{2(L_2 - L_4)}{\sqrt{L_3^2 - L_2^2 + 2L_4L_2 + L_4^2}}, \quad L_3^2 - L_2^2 + 2L_4L_2 + L_4^2 > 0 \tag{14}$$

which shows that when $L_2 = L_4$, the function has an extreme, in this case it is a maximum (**Figure** 7).

In the specific example when the length of the link $L_2 = 17$ [mm] the robot will move with maximum step S and as fast as possible under equal other conditions. In this case, the body and feet of the robot are raised to the same height $h_{Bmax} = h_{Smax}$. The driving mechanisms and the battery are located in the body of the robot, therefore the displaced masses in the two phases differ significantly. From the point of view of energy saving, it is more profitable to lift the body less, but this in turn leads to a reduction in velocity of the robot. An approach is applied in which the potential energy in the two phases of movement on flat terrain is equated.

The energy needed to lift the body during phase 1 is:

$$\mathrm{E_{p1} = m_1gh_{Bmax}} \tag{15}$$

The energy that the motor delivers in order to move the feet during phase 2 is:

$$E_{p2} = m_2gh_{Smax} \tag{16}$$

The equalized energies are as follows

$$E_{p1} = E_{p2} \rightarrow m_1h_{Bmax} = m_2h_{Smax} \tag{17}$$

From (17) and geometrical considerations from **Figure 6**, the following system is obtained

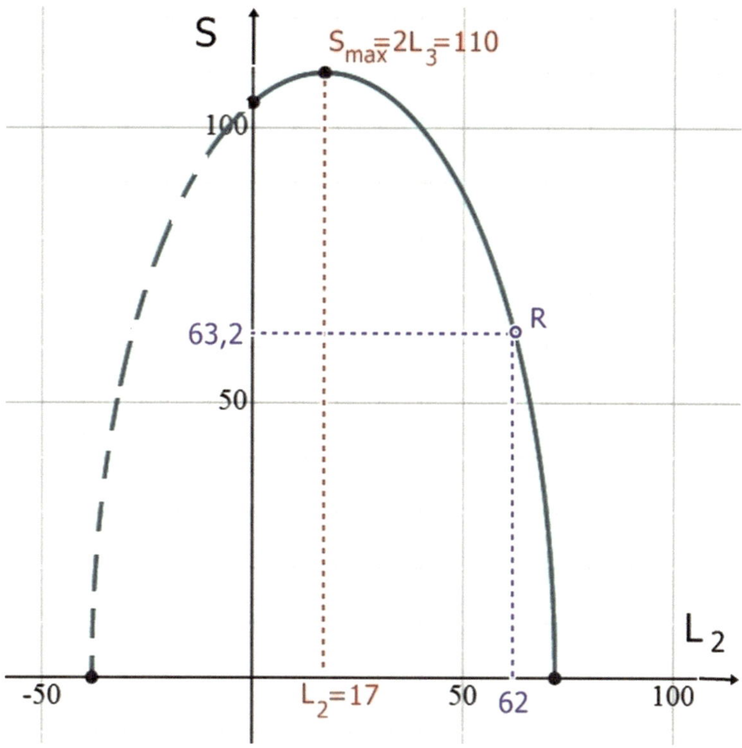

Figure 7.
Graph of the function for changing the step $S(L_2)$.

$$\begin{vmatrix} m_1 h_{Bmax} = m_2 h_{Smax} \\ h_{Bmax} + h_{Smax} = 2L_3 \end{vmatrix} \tag{18}$$

From (18), the height at which the body is lifted when the maximum potential energies, for the two phases are equalized, is as follows

$$h_{Bmax} = \frac{2L_3 m_2}{(m_1 + m_2)} \tag{19}$$

The weights of the 3D printed prototype are distributed in the two masses, respectively m_1 = 245 [g] and m_2 = 30 [g]. The maximum lifting height of the body h_{Bmax} = 12[mm] is obtained. At L_3 = 55 [mm] all parameters of the prototype are determined. These proportions of the lengths of the links not only improve the loading of the links and improve the distribution of energy in the two phases of movement on flat terrain, but also have a positive effect on overcoming high obstacles.

4. Passive adaptation to obstacles

When overcoming obstacles with height $h_b < h_O < h_S$, there are two ways to attack the obstacle: with the body (the round base {1} **Figure 1**) or with the feet.

If the height of the obstacle h_O is greater than the maximum lift of the body h_{Bmax}, the robot cannot climb on it during phase 1. In practice, it turns out that the robot can adapt to the obstacle and climb it by attacking it with the feet. It does not need special sensors and control algorithms. This process is illustrated in **Figure 8**. The robot body (round base) collides with the vertical section of the obstacle. Then there is a sliding of the feet on the horizontal terrain and the body is sliding on the vertical obstacle, the arm {3} performs a planer movement. It can be determined the instantaneous center of velocities of the arm {3} by taking into account the motion of points A and B from it. With respect to the absolute coordinate system, the instantaneous center of velocities of the link AB has coordinates:

$$\left| \begin{array}{l} X_Q = X_B - L_3 \sin(\alpha) \\ Y_Q = L_3 \cos(\alpha) \end{array} \right. \tag{20}$$

In this situation, the instantaneous velocity center of the arm jumps from point A_0 to point Q_1 and starts to move along an arc of a circle (**Figure 8**). The circle has radius L_3 and center $[-X_B,0]$ and its equation excluding the angle α, is derived from (20):

$$(X_Q + X_B)^2 + Y_Q^2 = L_3^2 \tag{21}$$

The relative instantaneous velocity center with respect to the coordinate system $[A_0,X',Y']$, and connected to the arm AB, is defined by the system of equations

$$\left| \begin{array}{l} X'_Q = L_3 \sin^2(\alpha) \\ Y'_Q = L_3 \sin(\alpha)\cos(\alpha) \end{array} \right. \rightarrow \left| \begin{array}{l} X'_Q = \frac{L_3}{2}(\cos(2\alpha) + 1) \\ Y'_Q = \frac{L_3}{2}\sin(2\alpha) \end{array} \right. \tag{22}$$

The relative trajectory of the instantaneous velocity center is also an arc of a circle, and its equation is derived from (22) after excluding α:

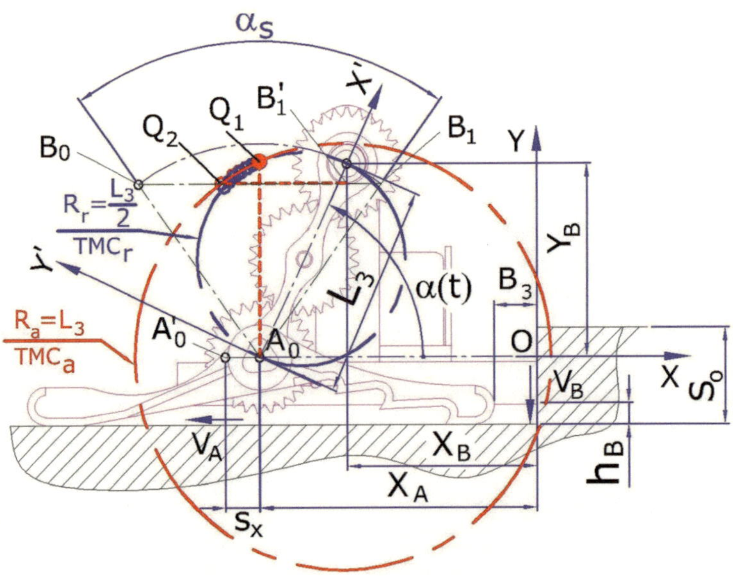

Figure 8.
Instantaneous velocity center and adaptive movements in case of collision between the robot's body and the obstacle.

$$\left(X'_Q - \frac{L_3}{2}\right)^2 + \left(Y'_Q\right)^2 = \left(\frac{L_3}{2}\right)^2 \tag{23}$$

Trajectory (TMC_a) is an arc from the red circle, which is described according to the absolute coordinate system OXY. Trajectory (TMC_r) is an arc from the blue circle, which is described with respect to the relative coordinate system, connected with the moving arm AB. In this situation the arms perform a planar movement and the body and feet of the robot are moving along X and Y axes respectively. When the body touches the ground, the instantaneous velocity center switches again and jumps to point B'_1. The feet begin to rotate and attack the obstacle. Video of the described passive adaptation is available from Video 5.

5. Overcoming obstacles

From the reasoning made so far, it can be seen that depending on the height of the obstacle, it is possible to overcome it when attacking with the body or to adapt to it and attack it with the feet. If the height of the obstacle h_0 is less than the maximum height reached by the feet h_{Smax}, several scenarios are possible:

- Overcoming the obstacle
- Rolling over of the robot
- Repeated sliding of the robot's feet and body on the obstacle, during which it cannot climb.

Figure 9 illustrates 5 stages when climbing an obstacle which differ in the elements of contact between the robot, the obstacle and the terrain.

1. The feet are in contact with the obstacle and the round base with the ground. Because of the rotation of the arm {3} sliding starts between the base and the

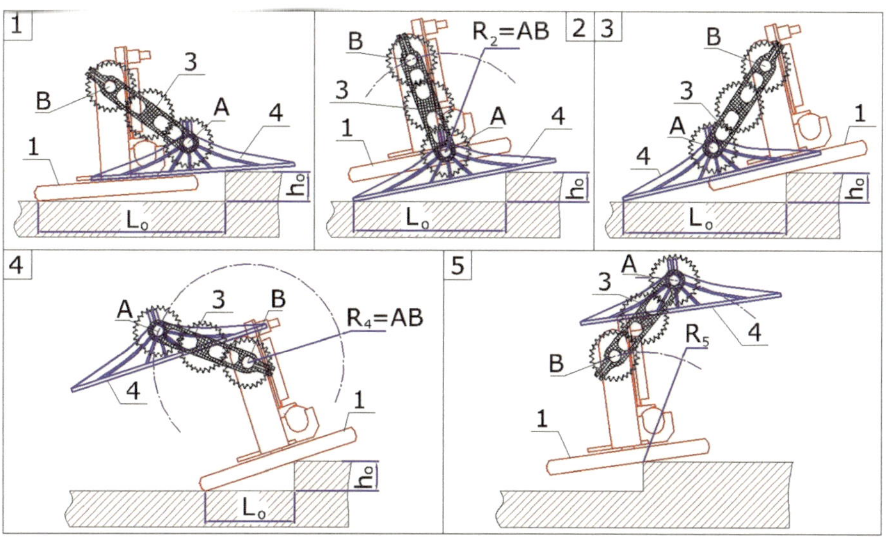

Figure 9.
Five consecutive stages when climbing an obstacle.

ground or between the feet and the obstacle. At this stage, it is good to ensure good traction between the feet and the obstacle, which will allow the robot to pull itself towards it.

2. The first stage ends when the feet are simultaneously in contact with both the obstacle and the terrain. The arm {3} rotates and moves the robot body. In this case, the feet are usually stationary, but for high obstacles, it is possible slipping to occur.

3. The round base of the robot has reached the obstacle. Due to the movement of the arm {3} a situation is reached in which the base is in contact with the obstacle and the feet in contact with the ground. The robot performs a planer movement in which there is sliding of the feet, the base or both on the terrain and the obstacle.

4. A configuration is reached in which the round base is in contact with both the terrain and the obstacle. The arms {3} rotate and they move the feet.

5. The center of gravity of the robot changes, shifting towards the obstacle. Depending on the height of the obstacle, the shape and the materials of the base, the masses of the links and the feet, it is possible to rotate point C around the edge of the obstacle in order to overcome the obstacle.

These stages are described in detail in [20, 21], where simulations and results of various experiments are presented.

6. Results from experiments with the 3D printed model

Although the 3D printed model is only a prototype, it can be used for experiments and useful conclusions can be drawn. 3D printing enables us to easily create and adjust prototypes. Already known key advantages of this technology are:

- Opportunity to create very complex external and internal areas

- Opportunity to create components with different density and internal infill structure

- Mixing multiple materials with different characteristics in the production of the same element (only with multi material printers).

Different versions of Big Foot were created. The first prototype used shafts with small diameters and with very small diameters and the transmission of torque is achieved by friction forces (**Figure 10** above). This leads to higher tension on the joints in the shafts and they have a tendency to slip.

In the second prototype (**Figure 10** below) some of the problems are solved. The feet shape is improved to secure better traction when overcoming obstacles. To avoid the shafts slipping, their diameter and contact surface is increased. A pin coupling is used (**Figure 10**, pos. 7), which is much more reliable, but leads to stress concentration. An innovative and patented coupling is successfully applied for the joint at point A, where the tension is highest [22] (**Figure 10** pos. 8a). It combines reliability of the contour joints with low levels of stress concentration. This coupling has the advantage to fix the foot to the shaft with constant orientation, which is

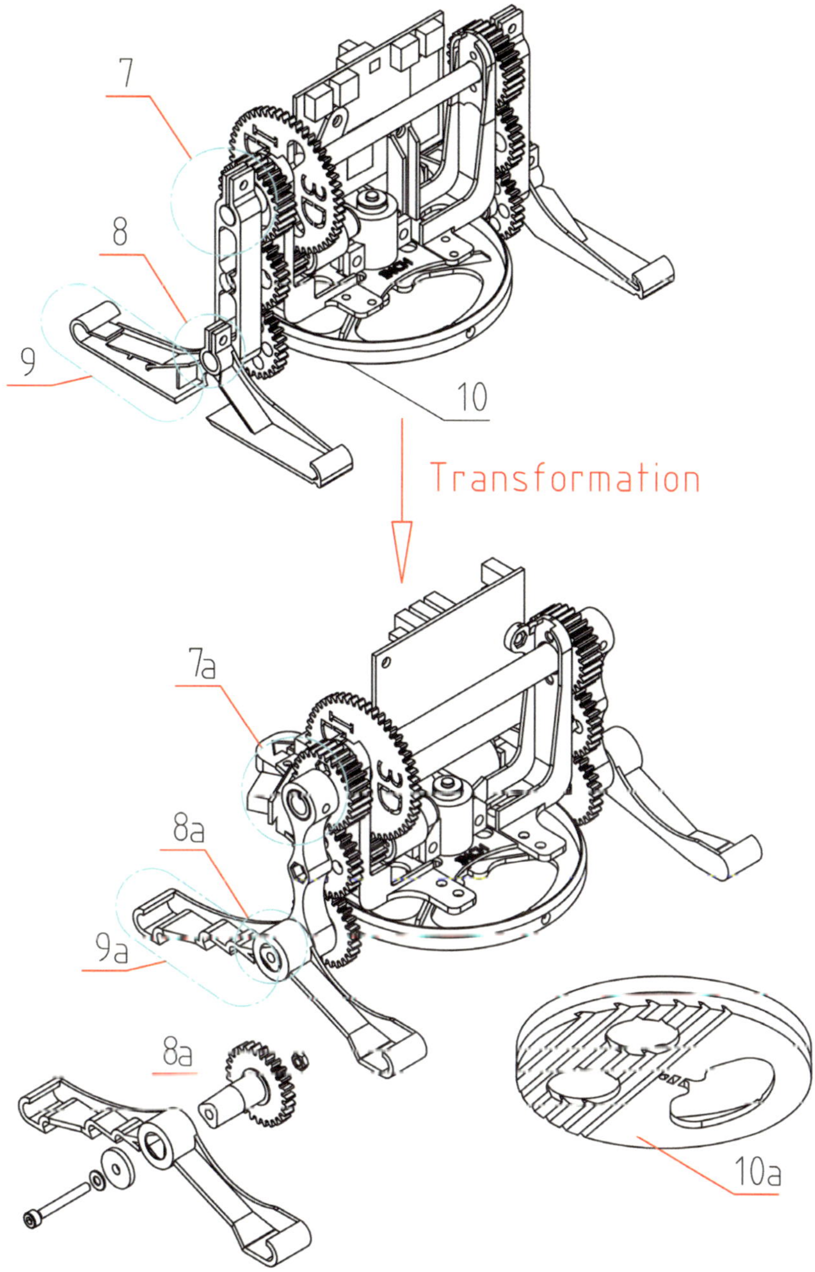

Figure 10.
Structural changes to big foot.

important for the proper functioning of the walking mechanism. The shape changes gradually from a reuleaux triangle to a circle. Such objects can be drawn using CAD products and can afterwards be 3D printed. The elements are held by a screw joint. To increase the max obstacle height, the shapes of the feet and the body's base (**Figure 10** pos. 9–10) are changed. The front jigged areas increase the traction and "pull" the robot, while the back edges are rounded which aids the "sliding". As there is no limit for the complexity of the 3D printed feet, the components can resemble nature more closely.

Figure 11 shows an analogy between a walrus and Big Foot when climbing. The movements share many common characteristics. A comparison with a walrus was

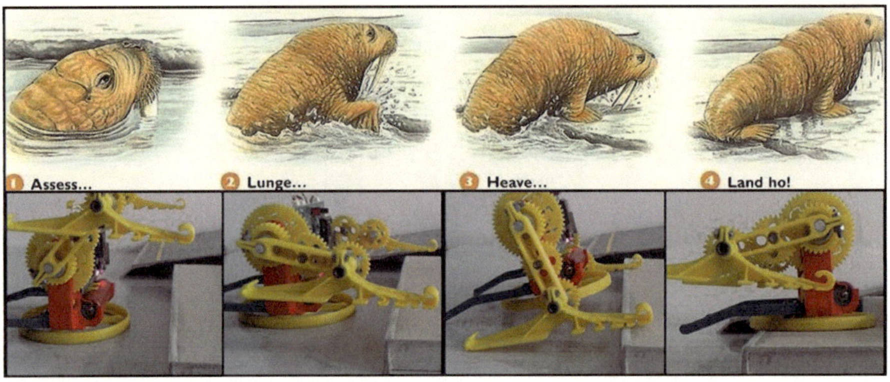

Figure 11.
3D printed feet with a complex shape, inspired by nature.

chosen because, like the animal, the robot pulls its heavy body on the obstacle with the help of sharp elements in its feet. Unlike perhaps all known animals, Big Foot can rotate its base and arms more than 360 degrees.

By increasing the width of the feet and the area of the round base the robot can walk on soft terrain (sand, snow, marsh) more easily.

Over 100 experiments for overcoming a higher obstacle were conducted. The same prototype was used, where only the feet {4} and the round base {1} were replaced. They were 3D printed and had different shapes. Two materials were used: PLA (Polylactic acid - most popular for 3D printing) and the flexible FIlaFLEX. The highest obstacle of 43 [mm] which was overcome can be seen here: Video 6; A detailed description of the results is available in [21]. After adding a "tail" to improve the balance of Big Foot, the maximum height was increased to 52 [mm]: Video 7; which corresponds to a coefficient K_{ro} = 0.41, see formula 8.

Experiments to overcome an obstacle with a maximum height are made with various mobile robots. Based on literary sources, [21, 23] their respective K_{ro}-indexes are defined and given in **Table 1**.

Using information from the literature, the coefficient K_{ro} can be determined for different mobile robots in **Table 1**. From the considered examples, it is seen that the highest value K_{ro} = 0.41 is associated with the Transformable-wheeled leg robot [21]. The Big Foot robot proposed in the present study has higher values of the K_{ro} index compared to the mobile robot [20] and the humanoid robot NAO. It can be noted that Big Foot manages to overcome this height by using only one of its two motors, while all other robots use several motors.

Robot	Height [mm]	Length [mm]	Maximum height of the obstacle [mm]	K_{ro} index	Number of motors
1. MSRox	290	830	100	0.20	2
2. NAO	640	160	70	0.22	25
3. Big Foot	88	182	52	0.41	2
4.Transformable-wheeled leg robot	180	390	120	0.45	3
5. Micro rover – Spacecat [23]	200	200	100	0.5	8

Table 1.
K_{ro} indices of different mobile robots.

7. Conclusion

An original design of a 3D printed walking robot based on minimalistic approach is presented. This idea is intended to inspire the design of useful robot structures in the future.

It is considered a design principle and determination of the proportions of the links, based on minimizing the energy during walking.

The kinematics of the robot are analyzed and the key stages for walking on flat terrain and climbing obstacles are given.

The principles of movement are considered and the robot's ability to adapt to obstacles due to the mechanical structure is highlighted. An algorithm is shown for calculating the change in the instantaneous velocity center of one link while the robot is adapting. This is a practical example of applying kinematic methods in robotics.

The main dependences for determining the torque loading of the motor when walking are given. The results of a study of the static conditions for overcoming an obstacle and experiments with a 3D printed model are discussed. Detailed studies and simulations are given in [20, 21]. 3D printing gives new opportunities to create unconventional structures, which can change the way robots are designed.

The results of experiments with different materials and shapes for the feet and the base of the robot are discussed. Thus is detected the maximum height of the obstacle that can be overcome. After additional design changes, this height is increased to 52 [mm]. An index K_{ro} is proposed which relates the robot's dimensions with the height of the obstacle it can overcome.

The results for overcoming an obstacle by different types of robots are ranked using the proposed index.

It is not easy to give definitive answers to the questions posed in the introduction. However, from the analysis of the literature and the results of this study it can be noted that:

If the number of degrees of freedom is less than two, the walking robot cannot be controlled to bypass obstacles. The idea proposed in [8] is debatable whether it can be characterized by one degree of freedom as it also uses controllable couplings. In addition, it is possible to realize the movements only sequentially.

The presented 3D printed model shows that it is possible to overcome obstacles by using a simple control system without sensors and feedback.

3D printing technology facilitates the creation of prototypes of the developed robots. It allows easy realization of links with complex shapes and connections between them.

Acknowledgements

These research findings are supported by the National Scientific Research Fund, Project N ДН17/10-12.12.2017.

Thanks

The author is grateful to his colleagues and in particular to PhD Bozhidar Naidenov for their help in creating the prototype, conducting the experiments and supporting this work.

Other declarations

All graphics and video material used are created by the author.

Video materials

All video materials referenced in the text are available at: https://bit.ly/3xSfT0K.

Author details

Ivan Chavdarov
University of Sofia "St Kliment Ohridski", and Institute of Robotics, Bulgarian Academy of Sciences, Sofia, Bulgaria

*Address all correspondence to: ivannc@uni-sofia.bg

References

[1] F. Rubio, F. Valero, C. Llopis-Albert, A review of mobile robots: Concepts, methods, theoretical framework, and applications, International Journal of Advanced Robotic Systems. March 2019. doi:10.1177/ 1729881419839596.

[2] U. Jahn, D. Heß, M. Stampa, A. Sutorma, C. Röhrig, P. Schulz and C. Wolff, A Taxonomy for Mobile Robots: Types, Applications, Capabilities, Implementations, Requirements, and Challenges, Robotics 2020, 9, 109; doi: 10.3390/robotics9040109.

[3] S. Caron, A. Kheddar, O. Tempier, Stair Climbing Stabilization of the HRP-4Humanoid Robot using Whole-body Admittance Control. IEEE International Conference on Robotics and Automation, May 2019, Montréal, France. .1109/ICRA.2019.8794348. hal-01875387v6.

[4] S. Faraji, H. Razavi, AJ. Ijspeert. Bipedal walking and push recovery with a stepping strategy based on time projection control. The International Journal of Robotics Research. 2019;38 (5):587-611. doi:10.1177/ 0278364919835606.

[5] M. Fevre, B. Goodwine, JP. Schmiedeler. Terrain-blind walking of planar underactuated bipeds via velocity decomposition-enhanced control. The International Journal of Robotics Research. 2019;38(10-11):1307-1323. doi:10.1177/0278364919870242.

[6] X. Luo and D. Xia, Impact Dynamics-Based Torso Control for Dynamic Walking Biped Robots, International Journal of Humanoid Robotics, Vol. 15, No. 1 (2018) 1850004 (25 pages), World Scienti¯c Publishing Company, DOI: 10.1142/S0219843618500044.

[7] G. Muscolo & C. Recchiuto, Flexible Structure and Wheeled Feet to Simplify Biped Locomotion of Humanoid Robots, International Journal of Humanoid Robotics, Vol. 13, No. 4 (2016) 1650030 (26 pages), World Scienti¯c Publishing Company, DOI: 10.1142/ S0219843616500304.

[8] A. Peidró, J. Gallego, L. Payá, J. Marín and O. Reinoso, Trajectory Analysis for the MASAR: A New Modular and Single-Actuator Robot, Robotics, 2019, 8(3), 78; https://doi.org/ 10.3390/robotics8030078.

[9] BGH. Smith, JR. Usherwood. Minimalist analogue robot discovers animal-like walking gaits. Bioinspir Biomim. 2020 Feb 7;15(2):026004. doi: 10.1088/1748-3190/ab654e. PMID: 31869827; PMCID: PMC7655146.

[10] K. Wang, D. Marsh, R. Saputra, D. Chappell, Z. Jiang, A. Raut, B. Kon, and P. Kormushev, Design and Control of SLIDER: An Ultra-lightweight, Knee-less, Low-cost Bipedal Walking Robot, Published in: 2020 IEEE/RSJ International Conference on Intelligent Robots and Systems (IROS), DOI: 10.1109/IROS45743.2020.9341143.

[11] B. Beigzadeh, M. Sabaapour, M. Yazdi, K. Raahemifar, From a 3D Passive Biped Walker to a 3D Passivity-Based Controlled Robot, International Journal of Humanoid Robotics, Vol. 15 (2018) 1850009 (27 pages), World Scienti Publishing Company, DOI: 10.1142/S0219843618500093.

[12] E. Corral, M.J. García, C. Castejon, J. Meneses and R. Gismeros, Dynamic Modeling of the Dissipative Contact and Friction Forces of a Passive Biped-Walking Robot, Appl. Sci. 2020, 10(7), 2342; https://doi.org/10.3390/a pp10072342.

[13] D. Mrozik, T. Mikolajczyk, L. Moldovan, D. Pimenov, Unconventional Drive System of a 3D Printed Wheeled

Mobile Robot, Procedia Manufacturing, Vol. 46, 2020, Pages 509-516, Publisher: Elsevier, https://doi.org/10.1016/j.promfg.2020.03.073.

[14] G. Stano, L. Arleo and G. Percoco, Additive Manufacturing for Soft Robotics: Design and Fabrication of Airtight, Monolithic Bending PneuNets with Embedded Air Connectors, Micromachines 2020, 11(5), 485; https://doi.org/10.3390/mi11050485.

[15] K. Tomczuk, Z. Malecha, 3D printed robotic arm with elements of artificial intelligence, Procedia Computer Science, Vol. 176, 2020, Pages 3741-3750, Publisher: Elsevier, https://doi.org/10.1016/j.procs.2020.09.013.

[16] M. Tyagi, G. Spinks, and E. Jager, Fully 3D printed soft microactuators forsoft microrobotics, Smart Materials and StructuresSmart Mater. Struct.29 (2020) 085032 (11pp) https://doi.org/10.1088/1361-665X/ab9f48.

[17] Y. Jia, X. Luo, B. Han, G. Liang, J. Zhao and Y. Zhao, Stability Criterion for Dynamic Gaits of Quadruped Robot, Appl. Sci. 2018, 8(12), 2381; https://doi.org/10.3390/app8122381.

[18] Q. Hao, Z. Wang, J. Wang and G. Chen, Stability-Guaranteed and High Terrain Adaptability Static Gait for Quadruped Robots, Sensors 2020, 20 (17), 4911; https://doi.org/10.3390/s20174911.

[19] S.Habibian, M. Dadvar, B. Peykari, A. Hosseini, M. Hossein Salehzadeh, and F. Najafi, Design and implementation of a maxi-sized mobile robot (Karo) for rescue missions, Journal reference: Robomech J 8, 1 (2021), Springer, DOI: 10.1186/s40648-020-00188-9.

[20] Chavdarov, I., & Naydenov, B. (2019). Design and kinematics of a 3-D printed walking robot "Big Foot", overcoming obstacles. International Journal of Advanced Robotic Systems. https://doi.org/10.1177/1729881419891329

[21] Chavdarov, I., Krastev, A., Naydenov, B., & Pavlova, G. (2020). Analysis and experiments with a 3D printed walking robot to improve climbing obstacle. International Journal of Advanced Robotic Systems. https://doi.org/10.1177/1729881420925282

[22] Chavdarov, I., 3D printed coupling between a shaft and a link, patent N 67072 B1, 2020, in Bulgarian, http://ir.bas.bg/fni2018/pdf/patent-67072b.pdf

[23] T. Sun, X. Xiang, W. Su, et al. A transformable wheel-leggedmobile robot: design, analysis and experiment. Robot AutomSyst 98; 2017: 30–41

Section 2

Humanoid Robots

Chapter 7

Communication and Interaction between Humanoid Robots and Humans

Arbnor Pajaziti, Xhevahir Bajrami and Gazmend Pula

Abstract

This paper deals with future robots that will be developed to assist and/or partially replace human activities that would provide for humans very much and frequently needed general-types of repetitive services for their daily tasks and engagements. As indeed the very name of humanoid robots intensely suggests, these engagements despite being routinely self-understood by implication as necessities of daily life, their frequency and repetitiveness, alongside other necessities of distributed elements of an increasingly intelligent daily environment, impose the need for deployment of various kinds of robots. It is to be assumed that there will be middle grounds between different types of humanoid robots, depending on the strength of their field of application. Collaborative robots that are conceived and intended to work i.e., collaborate safely with humans in a joint and shared workspace will expand and develop and be applied in increasingly diverse functions and working environments. Nowadays, intelligent robots are of course widely feasible and also increasingly available, but needless to say, even in the long run they will and cannot surpass the people in their creativity, their ability to learn in their differentiation, and maybe not even manage to catch up with all human complex requirements and needs. People will understandably continue to have a firm grip on the main switch.

Keywords: Artificial Intelligence, Humanoid Robots, Communication, Interaction, Kinect

1. Introduction

A variety of humanoid robots have been developed and researched in more than 500 research institutes and universities all over the world. Humanoid robots are robots that have a shape and form resembling that of humans including structural i.e., anatomical similarities such arms and legs as well as facial ones such heads containing eyes and mouths. However the most challenging and complex issue remains the development of the two-legged robots, reliable and capable enough to be meaningful partners to humans, in order to be able to perform actions that humans are capable of, but nevertheless are also needy of having them done all too frequently, especially those of a repetitive kind. Humanoid robots are thus intended to be used repetitive and laborious tasks, frequently more dangerous ones such as permanent inspection, necessary repetitive maintenance ones, especially of a repetitive nature or highly hazardous engagements that may emerge in various types of disasters areas as may be needed in case of emergencies in nuclear power plants.

Professional humanoid robots such as Honda and Sony have made significant advances that have enabled highly-capable humanoid robots [1]. Both companies invested more budget and manpower that enabled the design of small, powered joints that achieve power-to-weight performance unheard of in commercially available servomotors.

Humanoid Robots consists of Artificial Intelligence, sensors, mechatronics, and power. The prime task of humanoid robots nowadays consists in developing capabilities of recognizing visual expressions perceptions, and in view of that to enable addressing appropriately tasks of predicting what emotion the human is having by observing the visual facial expressions of humans. Therefore, all the humanoid robots need to have supplied are the data that will provide sufficient and appropriate information to have processed by means of which they will be able to perform and add to their available spectrum of learning and performing activities. The algorithms and other metrologies, such as Deep Learning, Neural Networks will be responsible to extract the features from these images provided to them.

All of these objectives for humanoid robots present a huge challenge and requirement for processing power and it is to be noted that it is not possible for a humanoid robot to use this kind of huge processing power alone. Therefore, the humanoid robots have to absorb, integrate and indeed capture the information from the surrounding environment and to deploy the cloud which the cloud will process further and feed it back to the humanoid robot.

Multisensory perception, cognition and, man–machine cooperation are technological fields of robotics that are being researched as key technologies today. Theses and processes from the research area of Artificial Intelligence (AI) are transferred to robotic systems. Considerable principles of biology serve as role models and recruiting potential for robotics. The highest possible kinematic form of the human body is reproduced with great accuracy. Humanoid, complex mechatronic systems inspired by biology are the prime new field of researching artificial intelligence. The humanoid robots are very interesting not only in their technology but also in its psychological aspects. The visions of science-fiction authors are gradually becoming a reality. That causes - in Europe and America more than in Japan - the number of respective critics of the issue i.e., topic to increase. Some already describe the horror scenarios of human-like robots who gain power over their masters and lead humanity to its downfall [2–4].

The transformative change to an information and knowledge society will steadily increase the acceptance of complex high-tech devices in the everyday environment of humans. According to the forecast of leading scientists, robots will therefore take on increasingly more and complex tasks in the private sphere of humans in the coming generations.

Fields of application for the species "Humanoid" is expected to continue to be focused on supporting humans in household activities, the elderly and nursing-type of care, or simply in some of entertainment aspects in families and households. They are intended to maintain or further improve the standard of living and amenities of the human environment.

2. Humanoid and collaborative robots replacing humans in work places

In the not-too-distant future robots are expected to handle a very considerable amount of all tasks and jobs, perhaps even a half of the actual contingent and thus contribute perhaps to leaving an "army" of people unemployed and perhaps getting concerned people considering this aspect very seriously. However, according to, certain views and analysis, in an alternative scenario, the same technologies that revolutionize certain important achievements as in the area of humanoid and collaborative robots, rather than reducing people's job opportunities, they contribute to raising living standards with new job opportunities not yet imagined and/or visualized clearly.

This is especially the case for collaborative robots, frequently referred to as Cobots, that apply the principle of robotic-type of automation of certain job types with repetitive activities and rather flexible and intelligent adaptations to work procedures. Thus the Cobots indeed contribute significantly in sequential automatic adaptation to job requirements in the common workspace by safely collaborating with human workers especially in job's repetitive and somewhat more menial job segments and tasks that are to be repeated routinely in endless and exhausting work cycles, potentially and especially in more cumbersome and dangerous ones. Due to the intense research and development in the area as well as in AI, by the next decade, the collaborative work and interaction between humans and humanoid robots is expected to become much more refined and indeed much more flexible.

The humanoid as well as collaborative robots will develop increasing capacities of cooperation in specifically designed environment. For example, Toyota is building such a specifically designed urban environment a future city in Shizuoka prefecture. Drones, autonomous buses, taxis and various types of humanoid/collaborative robots will be developed for wider-specter and higher-level collaboration and cooperation with humans. Such a specific environment will expectedly be conceived and designed from scratch in order to provide for such an intense and high-level human-Cobot effective functional coexistence especially in types of assembly factory jobs, delivery and security ones as well.

Indeed this interaction between the humanoid robots and humans is expected to become increasingly natural-felt for end-use consumers i.e., for the humans, both at the household and industrial production frameworks. This is highly probable as humanoid robots are expected to be able to increasingly absorb, capture, integrate and implement processed relevant information, especially as relates to the production environment as well as to the household one. One of the most utilized approaches is learning, interacting and implementing by imitation, by observing and integrating and functionalizing operator's behavioural patterns.

This can be expected with significant probability, especially due to the intense development of AI and related interphases that humanoid robots will be increasingly capable of predicting also human emotions in the forms of sounds of sighting and intricate and elaborate facial expressions of human that manifest the quality and intensity of related emotions and then mould it into a collaborative interaction.

Advances in artificial intelligence, or the ability of machines to learn by processing vast amounts of data, are doing a rethink i.e., reconsiderations of what is believed that only humans can do. Thus 2018 paper by the National Bureau of Economic Research found "a wide range of perspectives on public discourse, ranging from alarmist forecasts of mass unemployment caused by robots to optimistic forecasts of job creation."

Meta-learning which is implemented by reinforcement learning are the type of biologic models that are most commonly used in human-humanoid robot interaction.

However, with the virus pandemic catapulting the world deeper into the fourth industrial revolution, dubbed Industry 4.0 – the ongoing automation of traditional manufacturing and industrial practices, with artificial intelligence and robotics, under cover of "social distancing" which has caused an unwelcoming employment crisis for the working poor, with many of their jobs displaced by robots.

2.1 What jobs could be affected by Artificial Intelligence and robots?

Cashiers, clerks, cooks, waiters, receptionists, security guards, data analysts, tax-preparing personnel and truck drivers are among the jobs often cited as being the most susceptible to these concerns regarding advanced and enhanced development and application of automation.

Other professions that may be less vulnerable to these side-effects and concerns include surgeons, accountants and financial analysts.

Jobs that require repetitive activities to carry out tasks in a structured environment, mainly in production, are the first to be directly affected by automation.

Since 1980, the number of manufacturing workers in the US has decreased by a third, to about 13 million, while production has doubled.

The newer humanoid robots come equipped with "vision, mobility and learning abilities, doing more tasks". Sophisticated software can conduct phone conversations with clients, for example.

According to a study by the International Machinery Business Institute about 120 million workers in the world's 12 largest economies may need to be retrained in the next three years as a result of automation and Artificial Intelligence,

According to a this year's report from the Brookings Institution's Metropolitan Policy Program found that nearly 36 million Americans hold jobs with "high exposure" to being potentially negatively affected by automation.

It is considered and estimated that by 2030 many people will have to change jobs due to these side-effects of widely deployed automation processes.

2.2 What kind of jobs can be created due to of automation?

It has always been much easier to identify jobs at risk from technology than to anticipate the new types of jobs that can be created as a result of sweeping automation. Before the advent of the internet and smartphones, it would have been difficult to foresee the need for social media apps or specialists, much less the emergence of, for example, the "YouTube influencer" as a well-paid profession.

2.3 When will all this happen?

As it is being rather widely reported this has already begun. Thus according to the International Federation of Robotics Sales of professional service robots, those used for non-industrial functions like logistics, inspections, and maintenance have totalled some 271,000 units in 2018 accounting for a 61% increase as compared to the previous year 2017.

2.4 What can people do?

There is general agreement that human workers will require more education and skills to keep up with technological development and change and get accustomed and ready to change jobs and even professions more often than before, if and when required by the respective technological developments in the area of robotics and automation.

It is clear that there is hardly any rationale in, slowing down, stalling or preventing the expansion of automation processes to be applied for instance in manufacturing factories, by only considering potentially resulting and indeed likely job employment reductions and shedding a as the overall result might turn out to be negative and hence also counterproductive. Experts suggest that people should focus on the enhancement and automation of the production process and tasks being successfully completed rather than on the number of job employment opportunities, especially as automotive repetitive tasks provide more time available for much more creative and productive activities, which can result in the creation of even more new and creative job opportunities and profession that don't currently exist. According to the World Economic Forum some 133 million new positions and job opportunities might be created along these lines However, businesses shouldn't only set out to maximize profit with large scale deployment of automation processes and machines, but they must proactively seek new job opportunities and stimuli for their employees to help them advance their spectre of professional skill sets. Here's how to join the robot revolution.

2.5 What you'll pay?

If you have decided to buy a robot you have to search online at different sites such as auction sites, electronic stores and hobby shops, or seek out the components to build the robot type and shape based on your requirements. You will get different prices that depend on the number of sensors and motors, time of the processing speed, memory, battery life, and storage, etc.

For example, the Walker robot shown in **Figure 1** is an intelligent humanoid robot to ease your everyday household work and making life easier, smarter and more convenient. It has two seven degrees of freedom robotic arms which provide a wide range of arm movements, flexible manipulation and obstacle avoidance by using visual and first sensor. It can also maintain its body language while moving and carrying objects. Using gait planning and control it can adapt to complex surfaces and walk on any surface required easily by using advanced control algorithms and thus it can maintain stable control of its hands and arms while swiftly moving through the surrounding environment. With a new vision navigation system, this robot can recognize contour colour depth and others without any visual aids.

The first Williams- robot type worked with the now-discontinued Aibo-robot type as shown in **Figure 2**, a dog-like robot manufactured by Sony between 1999 and 2005.

Aibo is a robotic pet that brings warmth with lovable behaviour and delights your everyday life. It is equipped with a 64-bit quad-core CPU which can deliver fast performance and interaction to provide you real dog-like experience. This robot

Figure 1.
Walker robot [5].

has one main camera and another slanted camera in which animals need to memorize up to 200 different interactions and can recognize and respond appropriately to them. The Aibo-robot is also equipped with six sensors, a motion detector and a lie detector which enables the robot to detect obstacles and move flawlessly around the house. This robot has also four microphones, thus being able to hear and respond accurately to your voice commands. You can get this robot at around three thousand dollars and they are available online.

The Temi robot, as shown in **Figure 3**, is the first robot that interacts with humans while providing a flawless connection between devices and your loved ones. It is equipped with a navigational robot system, 360-degree Lidar, true depth camera, RGB camera, 5 proximity sensors and real-time sensor fusion which analyses data and ensures autonomous navigation through a 3D mapping path, planning obstacle avoidance using detection and tracking its features at 10.1 inches per second. LCD touch colour display with a pixel density up to 225 PPI, comprising a brushless DC motor and planetary gear with which it can autonomously track the face and tilt the screen with accuracy you to interact with a robot with the clarity it has a 13-megapixel high resolution which can record thousands of ATB videos at 30 FPS while providing two-way live conversation with their loved ones. Temi-robot has 20 Watt speakers with high fidelity equalizers which provide the best quality music. It also has four omni-directional i.e., all-directional digital mics with real-time localization, in order to provide the best audio call experience. With built-in Alexa, one can command the TV to play music, place calls, check the weather and even control smartphone devices without leaving your comfort. Temi is a personal robot that you can order and get online for a price of some1500 dollars.

2.6 Speech recognition

To create a humanoid robot to enable speech recognition one has to use different hardware and software elements. These elements are as follows: Python

Figure 2.
Aibo robot [6].

Figure 3.
Temi robot [7].

programming language, different AI packages like speech-recognition and chatter-pot that need to be integrated into a pocket PC such as Raspberry Pi. Nowadays, humanoid robots can recognize the words of multiple people speaking simultaneously, can get certain information from the internet and so on. Certain types of them can be used in halls and offices and can communicate with many people [8].

Nao humanoid robot is also able to see, talk and hear. Nao can also naturally interact with humans. A shown in **Figure 4**, it has 4 built-in microphones and loudspeakers, 2 cameras. Nao can learn and adapt to almost every interaction and becomes more and more intelligent with time and empirics i.e., experience. He remembers answers and content and can immediately use them again in similar situations. It acquired its skills through a programming interface to IBM Watson's Language, Vision, Speech and Data APIs. These present almost endless possibilities for further development.

The Sophia from Hanson Robotics shown in **Figure 5** is a good example of how the AI is implemented in humanoid robots. Within its Robot Intelligent System it has some unstructured language learning as well as statistical natural language learning and natural language generation, but for some answers she might go to the Web, and some of the answers, might go to natural language learning. However some answers might enter into its robotic personality and hence it can behave similar to a human. The above-mentioned components that are implemented in the hardware and software are not distinct things, the real cracks of intelligence are in fact how they come and interact together to form an entire architectural organism as shown in **Figure 5**. The AI algorithms are included in Humanoid Robots for reasoning (logics), learning, perception and interaction, all of which as a whole inter-operates together in a complex way of communication and interaction with humans.

Figure 4.
Nao robot [9].

Figure 5.
Sophia Intelligent Robot [10].

3. Humanoid robot testing

TRI – the Toyota Research Institute was founded in 2016. Their role is to perform research, identify and create new capabilities Toyota intends to have in the future [11]. Toyota is trying to approach the future from a human-centred perspective with the goal of facilitating and bringing significant amount of fulfilment and happiness preferable to a majority of people of all walks of life on a basic i.e. fundamental level. This pursuit is based on a powerful idea that is contained even in some prominent constitutions that each person's life should strive towards happiness, meaning and purpose. In Japan, this is called Ikigai. Studies of Ikigai teach people that they feel most fulfilment when their lives incorporate work that they love and help society to enable more people to achieve their Ikigai. They pursue new forms of automation in society with a human touch to develop capabilities that amplify, rather than replace human ability. This is Toyota's historic philosophy of Jidoka, an idea that embraces the concept of Artificial Intelligence (AI), in other words, the human, and the machine work together in order to do something better than if either one of them could do on their own.

They are currently pursuing this vision in four research areas: Robotics, Automated Driving, Accelerated Materials Design and Discovery, and Machine Assisted Cognition [11].

TRI vision and mission are focused on solving the problem of how technologies can enhance and ease the human experience bringing forth a higher quality of life, independence and happiness. TRI envisions a future where Toyota products can improve the quality of life for societies around the world with an outstanding performance and contribution, their mission being the development of automated driving, robotics and other human enhancement and amplification technology from Toyota. Technological capabilities that will help people navigate safely from their kitchen to their living rooms, or safely across town, and most importantly, by providing this kind of human amplification technology, they hope to make the quality of life for everyone much better [11].

A growing number of Japanese businesses are testing robots as a viable solution to the country's shrinking workforce. They're popping up in stores, banks and soon are expected also in hotels. Bank of Tokyo-Mitsubishi UFJ is testing Nao, a robotic client service that answers basic questions and is designed to speak 19 languages. Multilingual polyglot robotics has been planned to serve foreign clients during the Tokyo 2020 Olympics.

By the time, the bank hopes to have even more robots on its staff. Pepper is a humanoid robot talking to clients. A humanoid has human-like features, for example, arms, legs as well as a head - but it is designed to look like a robot. Producer Softbank hopes Pepper will be a family robot, as in the Jetsons cartoons.

A hotel planned to open at Huis Ten Bosch Amusement Park in Nagasaki this summer aims to have 10 robots as staff members and soon to increase the number to more than 90 percent of hotel services being provided by robots as shown in **Figure 6**.

Today's innovation may be the necessity of tomorrow. Japan has an aging population that has fuelled heated debates about the involvement of robots in the state's workforce. A survey by home service operator Orix Living found that more seniors feel comfortable being nursed by a robot than when receiving services from a foreign nurse. The number of elderly citizens in Japan is steadily increasing, thus bringing about a real need for humanoid service robots to help them out in dealing and taking care of various home tasks.

In a country where the population is shrinking due to various reasons, where the workforce is shrinking and there is considerable resistance to an influx of immigrants as in Japan, it appears that robots may play a very big role in their future [13].

3.1 Pepper robot understands the emotions

One group that seems to want to embrace robots are elderly citizens of Japan who are cared for by robots. Using emotion recognition functions, the Pepper robot, released in February 2015, can understand and respond to people who joke, dance, and even make rep music in the Japanese language, see **Figure 7**. Pepper robot is

Figure 6.
Robot chief in preparing the pancakes [12].

Figure 7.
Pepper robot at the working place [14].

1.20 meters tall, has been designed by Softbank Robotics and can handle a conversation with people.

It can analyse human expressions, voice tones and gestures, thereby enabling them to respond. This type of robot can serve for education, health and entertainment purposes, its primary purpose however being not hard work, but home entertainment or shopping.

Pepper is the ideal robot for a family and can very quickly become the family of those individuals living alone and feeling lonely in their households, as it makes the elderly feel very comfortable in their interactions with these types of humanoid robots. In this context it should be mentioned that the Japanese society is prone to a friendly approach in its relations to robots in general and humanoid robots in particular. This is related also to their history and especially their world famous manga books.

4. The gesture-based remote human-robot using Kinect

4.1 Structure of the control system

Kinect is a line of motion sensing input device signals produced by Microsoft. Initially, the Kinect was developed as a gaming accessory for Xbox 360 and Xbox One video game consoles and Microsoft Windows PCs shown in **Figure 8**.

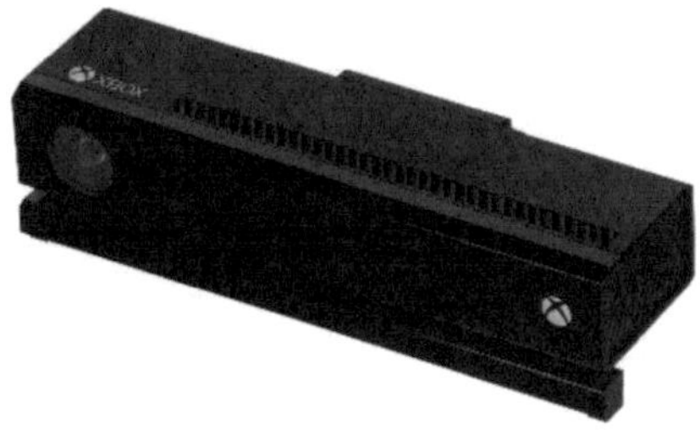

Figure 8.
Kinect sensor [15].

Microsoft Kinect sensor is comprised of three sensors: an infrared projector, an infrared camera and RGB camera to capture high resolution 3D images. The Kinect sensor is a popular sensor for robotics due to the advanced capabilities it offers for the human-robot interaction. Microsoft Kinect sensor is a major innovation in robotics.

With the use of dedicated software, users can easily control the movements of a robot by using an Xbox Kinect and their bodies to dictate and instruct the desired modalities.

4.2 Software description

A Microsoft Kinect v2 camera is used to track human motion using skeleton tracking. This technique has some limitations on tracking particular motions, especially motions of the palm of the hand that cannot yet be recognized. For example, the motion primitive "close hand" can be commanded while remotely operating the arm to hover over the grasping position. An online tracking system has been developed to control the arm of a Bioloid robot using Kinect sensor. The task of this work was hand-guiding robot arms using Microsoft Kinect v2. This objective has been achieved using a Kinect v2 and a Bioloid robot, which is a humanoid robot with 18 degrees of freedom (DOF) in total. The joint motions of the operator's arms and legs in the real world captured by a Kinect camera can be transferred into the workspace mathematically via forward and inverse kinematics, realistically through data-based UDP connection between the robot and Kinect sensor. The user assumes a specific pose to initiate a skeletal tracking. After the tracking begins, the user can start controlling the robot. After turning the motors on, the user can operate the robot remotely. The initial location of the user becomes the origin of the control coordinate system.

5. Connecting the Bioloid robot with Kinect

This system consists of both hardware and software. The way it functions is by capturing the user gestures, processing them and sending the processed signal further to the humanoid robot. Initially, the user makes a certain body gesture maintaining it for a short period of time [16].

The Kinect sensor is then used to capture the depth image of the user and recognizes the gesture by tracking the user's skeleton. This stage is called the image capturing stage. The depth image captured is processed into a computer in order to obtain an approximation of the positions of each body joint. In the gesture recognition stage, the angle between some of the body joints are then calculated and used as features for gesture classification. Once the correct gesture is recognized robot will then execute the motion correlating it with the recognized gesture [16]. The robot receives the command via a wireless interface. A built in mechanism is also embedded in by the PC with additional stability control to maintain the balance while moving and giving it the ability to get back up from potential falls autonomously [17].

The hardware used for this system are: Kinect sensor, PC, humanoid robot-kit, along with other additional tools. The Kinect used in this research is a depth imaging camera originally used for entertainment and gaming for Xbox game console made by Microsoft Corp. The humanoid robot kit used is Bioloid Premium Kit.

Above is presented the Kinect v2 connection method with the Bioloid robot using V-Rep simulation software.

The robot simulator V-REP with the integrated development environment is based on a distributed control architecture: each object/model can be individually controlled via an embedded script, a plug-in, a ROS or BlueZero node, a remote API client, or a custom solution. This makes V-REP very versatile and ideal for multi-robotic applications. Controllers can be written in C/C++, Python, Java, Lua, Matlab, or Octave. V-REP is used for fast algorithm development, factory automation simulations, fast prototyping, and verification, robotics-related education, remote monitoring, safety double-checking, etc. [18–20]·

A direct connection for the human gait parameters using Kinect camera that is capable of providing human body tracking in real-time is shown in **Figures 9–11**. The position of the hands is then continuously updated and relayed to the robot, which moves towards the indicated position.

5.1 Pseudo-code for communication with a humanoid robot

The pseudo-code for connecting Kinect to MatLab and v-Rep software is given as follows:

Algorithm 1

Initializing parameters $\leftarrow$ *Neck, Head, Right Leg, Left Leg, Right Hand, Left Hand&&Spine* $\in$ *SkeletonConnectionMAP*

Insert variables: **if**

$$SkeletonConnectionMAP \in \begin{bmatrix} 3 & \cdots & 15 \\ \vdots & \ddots & \vdots \\ 12 & \cdots & 24 \end{bmatrix}_{MxN}$$

end

Insert variables: VREP API

$\mathbf{vrep} \leftarrow remApi\{vrep.simx \in \forall;$

$clientID \in remApi||true;$

$\mathbf{if} \leftarrow clientID > -1;$

$\mathbf{disp}(\mathbf{vrep} \in API);$

end

Create Right Arm :

$$\mathbf{return} \leftarrow vrep.simx.Object \begin{bmatrix} 1 & \cdots & 3 \\ \vdots & \ddots & \vdots \\ 4 & \cdots & 6 \end{bmatrix} \in clientID\&\&ART(n:m);$$

Create color and depth videoinput objects :

$\mathbf{colorVid} \leftarrow input \in kinect_1;$

$\mathbf{depthVid} \leftarrow input \in kinect_2;$

$\mathbf{depthSource} \leftarrow Frame, Trigger \in depthVid;$

$\mathbf{himg} \leftarrow figure;$

while

$trigger \in depthVid\&\&colorVid;$

$colorIMG \leftarrow getDATA \in colorVid;$

$$metaDATA \leftarrow getDATA \in depthVid;$$

if

$trackedBODIES \leftarrow find \in metaDATA;$

$trackedBODIES \leftarrow find \in metaDATA;$

$jointCoordinates \leftarrow metaDATA \in JointPOSITIONS;$

$ColorJointIndices \leftarrow metaDATA \in JointINDICES;$

$robotArmControl\ \{$

$image \leftarrow colorImg\ (\langle : | : | : \rangle, 1);$

$nBODIES \leftarrow length \in trackedBODIES;$

for $i = 1 : 25;$

for body $= 1 : nBODIES;$

end

end

end

end

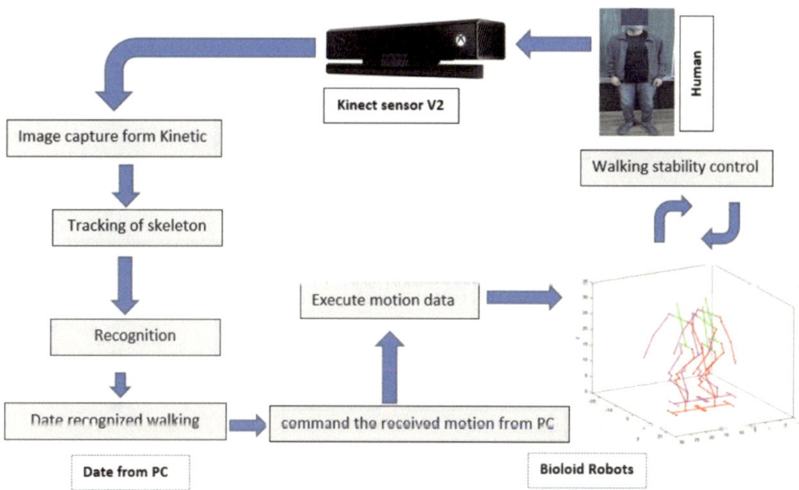

Figure 9.
Interaction between Bioloid Robots and Humans control architecture.

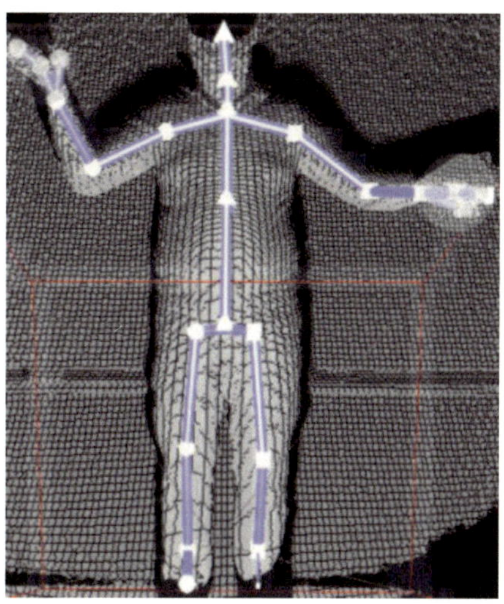

Figure 10.
Extraction of arm reference points.

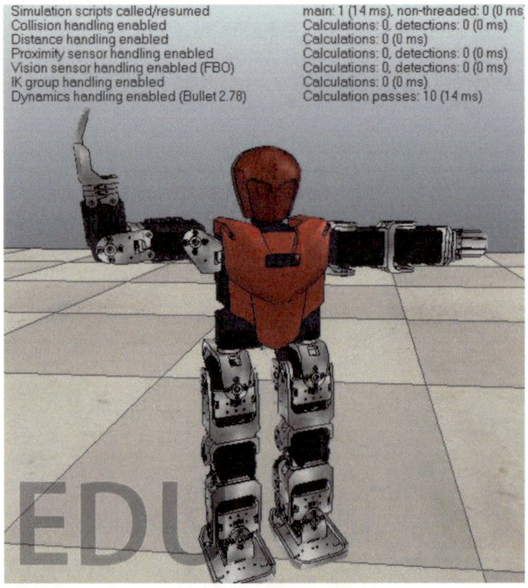

Figure 11.
Arm movement after parameter setting.

6. Conclusions

Regarding the application and future development of Humanoid Robots the following conclusions could be summarized as presented below:

- dependency on the expected wider-scale deployment and dependency on humanoid robots in daily life of citizens is likely to expand significantly and thus influence increasingly daily routines of their everyday life.

- If accepted on a wider scale by people, in the very near future these humanoid robots could become as important as computers currently are already.

- It could potentially be expected that in the foreseeable future human communication and interaction with a humanoid robots is likely to become increasingly similar to inter-human communication.

- The authors have independently developed and presented the Kinect v2 connection method with the Bioloid robot using V-Rep simulation software.

Thus the Cobots indeed contribute significantly in sequential automatic adaptation to job requirements in the common workspace by safely collaborating with human workers especially in job's repetitive and somewhat more menial job segments and tasks that are to be repeated routinely in endless and exhausting work cycles, potentially and especially in more cumbersome and dangerous ones. Due to the intense research and development in the area as well as in AI, by the next decade, the collaborative work and interaction between humans and humanoid robots is expected to become much more refined and indeed much more flexible. The humanoid as well as collaborative robots will develop increasing capacities of cooperation in specifically designed environment.

Author details

Arbnor Pajaziti[1], Xhevahir Bajrami[1*] and Gazmend Pula[2]

1 University of Prishtina, Faculty of Mechanical Engineering-Department of Mechatronics, Kosovo

2 University of Prishtina, Faculty of Electrical and Computer Engineering, Kosovo

*Address all correspondence to: xhevahir.bajrami@uni-pr.edu

References

[1] Siegwart R., Nourbakhsh I. R. Introduction to Autonomous Mobile Robots, MIT, 2004.

[2] Schraft R. D., Haegele M., Wegener K. *Service Roboter Visionen*. Hanser, 2004.

[3] Schraft R. D., Schmierer G. *Service Robots*. A K Peters Ltd, 2000.

[4] https://www.mathworks.com/help/supportpkg/kinectforwindowsruntime

[5] https://spectrum.ieee.org/automaton/robotics/humanoids/ubtech-upgrades-walker-humanoid-robot

[6] https://www.thehindubusinessline.com/info-tech/sony-revives-pet-ai-project-with-updated-aibo-robot-dog/article9936932.ece

[7] https://nocamels.com/2020/03/personal-ai-robot-israel-temi-robotemi-coronavirus/

[8] Corke P., Sukkarieh S., *Field and Service Robotics*: Results of the 5th International Conference (Springer Tracts in Advanced Robotics), 2006.

[9] Shamsuddin, Syamimi, et al. "Humanoid robot NAO: Review of control and motion exploration." 2011 IEEE international conference on Control System, Computing and Engineering. IEEE, 2011.

[10] https://news.un.org/en/story/2017/10/568292-un-robot-sophia-joins-meeting-artificial-intelligence-and-sustainable

[11] https://www.tri.global/

[12] https://www.dailymail.co.uk/sciencetech/article-4180896/Robot-chef-prepares-pancakes-robot-kingdom.html

[13] Joo-Ho Lee, Guido Appenzeller and Hideki Hashimoto. *Field and Service Robotics*, Section in book by Alexander Zelinsky (Editor). Springer Verlag, 1998.

[14] https://www.softbankrobotics.com/emea/en/pepper-workspaces-ga

[15] https://www.tejar.pk/microsoft-kinect-sensor-for-xbox-one

[16] M. A. Ma'sum, M. S. Alvissalim, F. Sanjaya, Andros and W. Jatmiko, "Body gesture based control system for humanoid robot," 2012 International Conference on Advanced Computer Science and Information Systems (ICACSIS), Depok, 2012, pp. 275–280.

[17] Bajrami, X., & Likaj, R. (2017). Dynamic Modeling and Simulation of Humanoid Robot. LAP LAMBERT Academic Publishing.

[18] https://github.com/LAB08-SBC/Bioloid-Java-Simulator

[19] http://www.co ppeliarobotics.com/

[20] Pajaziti, A., Bajrami, X., Shala, A., & Likaj, R. (2018, September). Dynamic walking experiments for humanoid robot. In Proceedings of SAI Intelligent Systems Conference (pp. 866–880). Springer, Cham.

Chapter 8

Safe Adaptive Trajectory Tracking Control of Robot for Human-Robot Interaction Using Barrier Function Transformation

Iman Salehi, Ghananeel Rotithor and Ashwin Dani

Abstract

In this chapter, safety methods in human-robot (HR) interaction/collaboration are presented. Ensuring the safety of humans, objects, or even the robot itself in the robot's operating environment is one of the crucial aspects of collaborative robotics. Since there are limited ways of controlling the behavior of humans, e.g., by placing physical barriers, shaping the behavior of the robot is a feasible option. The chapter discusses current methods of placing barriers for human safety in an industrial setting and novel methods of placing virtual barriers by designing robot controllers using barrier transformation. The concepts of barrier functions (BFs), control barrier functions (CBFs), and barrier transformations are reviewed. The barrier transformation concept is used to design an adaptive trajectory tracking controller for the robot such that the robot does not cross the virtual barriers. The designed controller is tested in simulations. Future directions of safety technology in human-robot collaboration are presented.

Keywords: Safety, Barrier Transformation, Trajectory Tracking Control, Human-robot collaboration, Safe adaptive control

1. Introduction

In many robotics and other engineering applications, maintaining system states within a prescribed bound is essential to satisfy the system safety property. For example, in a manufacturing collaborative robotics context, it is crucial for the robot to satisfy requirements, such as trajectory boundedness and to safely carry out its operations [1–3]. In medical robotics context, when a robot is interacting with a person, the person undergoing surgery cannot move so the robot must stay within virtual barriers in 3D space so that it does not harm the person. See [4] for an example cobot architecture. A review of recent methods for safe human-robot (HR) interaction methods is presented in [5].

In this chapter, safety in the context of HR collaboration is defined such that the robot does not cross over a prescribed physical space where humans or other robots are operating or the robot does not cross joint or the task space limits when the robot is collaborating with the person. The violation of constraints can lead to severe degradation of the robot's performance, unsafe behavior, and sometimes failure of the robot's components. In collaborative robotics applications such as collaborative manipulation [6], collaborative construction [7], teleoperations [8, 9], for human-in-the-loop control applications [10], or distributed multi-robot control applications [11–13], restricting the motion of the robot to a constrained configuration or task space is essential. Safe HR collaboration/interaction is also important for introducing robot factory co-workers in manufacturing automation [14, 15], developing robotic assistants for astronauts [5], for assistive robotics [16–20].

The literature related to the control of the robot in HR interaction focuses on designing impedance control laws [16, 21–23] or admittance control laws [23, 24] for adapting the interaction forces exerted by the human on the robot, when the robot is physically interacting with the human. In [24], an admittance controller is designed which takes inputs from human actions to achieve safer HR interaction. In [25], a physical human-robot interaction in the context of bikebot is presented. In [21], a an adaptive impedance controller for HR interaction is developed that is based on the NN model of the human intention. In [26], a controller is developed for human robot handover interaction based on dynamic movement primitives. In these examples, the human is physically interacting with the robot.

Other studies in the literature address the problem of robot/autonomous system control to avoid running into humans by modeling them as obstacles [27–29]. Most of these studies view the problem as a collision avoidance problem and solve the collision avoidance using potential field approach [30]. These control actions are purely reactionary in nature [31]. To achieve pro-activeness, studies in literature have designed controllers and motion planners that incorporate the probabilistic information about the possible intentions of human actions [32–34]. When humans and robots collaborate, inference of the person's intentions or robot's intentions improves the overall performance of the collaborative task [35]. Many studies in the literature have focused on designing scheduling and planning algorithms. In [36], a stochastic trajectory optimizer for motion planning is used for planning robot arm motion based on human intentions. In [37], scheduling, planning and control algorithms are presented that adapt to the changing preferences of a human co-worker, while providing strong guarantees for synchronization and timing of activities. In [38], new hierarchical planners based on Hierarchical Goal Networks are developed for assembly planning in human-robot team.

In the context of control architecture design for human-in-the-loop systems, adaptive controllers are presented using the inner-outer loop control structure in [10]. Stability studies of human-in-the-loop telerobotics with time-delay is presented in [39]. However, these studies do not explicitly consider safety aspects of the human-in-the-loop systems. Providing safety guarantees on the learned controller of machine/robot is typically achieved by adjusting the reference command using a pre-filter called a reference governor [40, 41] or by using optimal control under uncertainty in a differential game setting.

For keeping the robot state bounded in a prescribed bound saturated controllers can be used [42, 43]. Barrier function (BF) is a commonly used approach to certify the forward invariance of a closed set with respect to a system model, which can be used to examine the system's safety property [44, 45]. There are two candidates to construct BFs, namely, Reciprocal BFs and Zeroing BFs. The Reciprocal BFs can be of inverse-type and logarithmic-type. Extensions of BFs to controlled systems called as control Barrier Functions (CBF) have also been developed in the literature [46, 47]. Applications of BFs or CBFs in many autonomous robotic systems, such as robot manipulators, autonomous vehicles, and walking robots, are shown in [48–50]. In [47, 49, 51], BFs were successfully applied to dynamical systems where ensuring safety conditions are critical. In [51], time-varying BFs and CBFs for avoiding moving and static obstacles are derived, and their application to flying quadcopter is shown which avoids unsafe obstacle regions. Robustness properties of the CBFs are studied in [46], which shows that if a perturbation (or model error) makes it impossible to satisfy the invariance condition for a reciprocal barrier function, then the solution of the model must cease to exist because the control input becomes unbounded. For the Zeroing CBFs, Input-to-State stability (ISS) result holds in the presence of model uncertainties. A concept of exponential BFs and CBFs is introduced in [52]. The method of CBFs is extended to position-based constraints with relative degree 2 in [53] to address the safety constraints for systems with a higher relative degree. Furthermore, a backstepping based design method to design CBFs with a higher relative degree is also introduced. However, achieving a backstepping-based CBF design for systems with a higher relative degree is challenging. In [52], a concept of exponential CBFs is introduced that can handle state-dependent constraints for systems with a higher relative degree. In [54], a safety aware RL framework using BFs is proposed.

Barrier Lyapunov function (BLF) is another method that is used for the control of nonlinear systems when the outputs and states have upper and lower bound constraints (cf. [55, 56]). The BLF is constructed such that its value grows to infinity whenever its argument approaches the bounds. In [55, 57], an adaptive controller is developed using BLF defined over the output tracking error for single-input and single-output (SISO) nonlinear systems in a strict-feedback form. The controller works when the constraints are either constant or time-varying output constraints. An extension to output tracking with partial state constraints is developed in [58]. Using a similar BLF, in [59], an adaptive neural network with full-state feedback control that uses a Moore-Penrose pseudoinverse term in the control law design is developed for an uncertain robot dynamics with output constraints, and the signals of the closed-loop systems are proven to be semi-global uniformly ultimately bounded (SGUUB). In [54, 60], a BLF method that uses reinforcement learning (RL) is developed for a state regulation problem of a SISO nonlinear systems in the Brunovsky form with full-state and control input constraints.

Designing safe controllers using learning based control methods are also presented in the literature. For example, in [61], a safe, online, model-free approach to path planning with Q-learning is discussed. A general safety framework for learning-based control using reachability analysis is presented in [62]. In [63], a receding horizon safe path planning approach using mixed integer linear programming (MILP) is presented. Safe trajectory generation for autonomous operation of spacecraft using convex optimization formulation is proposed in [64]. When the region is non-convex, successive convexification can be performed [65]. A detailed survey and tutorial of $\mathcal{L}_1$ adaptive control architecture for safety critical systems is presented in [66].

In this chapter, barrier function transformation, presented in [67], is used to design a safe adaptive trajectory tracking controller for the robot using Euler–Lagrange (EL) system. The safe adaptive trajectory tracking control architecture of a robot system presented in this chapter is shown in **Figure 1**. Full state constraints are used while designing the torque control law. A gradient parameter update law is designed along with projection laws to keep the parameter estimates bounded. A Lyapunov-based stability analysis is presented which concludes semi-global uniformly ultimately bounded tracking result. Simulations studies are conducted using 2-link robot such that the tracking controller does not cross the bounds placed on the joint angles of the robot leading to a desired end-effector motion within a certain bounds. In addition to the control design and its testing in simulation, the chapter presents a review of standard techniques of designing safe robot controllers using BFs and CBFs, followed by a review of Barrier transformations which is used to design adaptive robot controller of EL robot system in this chapter. Future

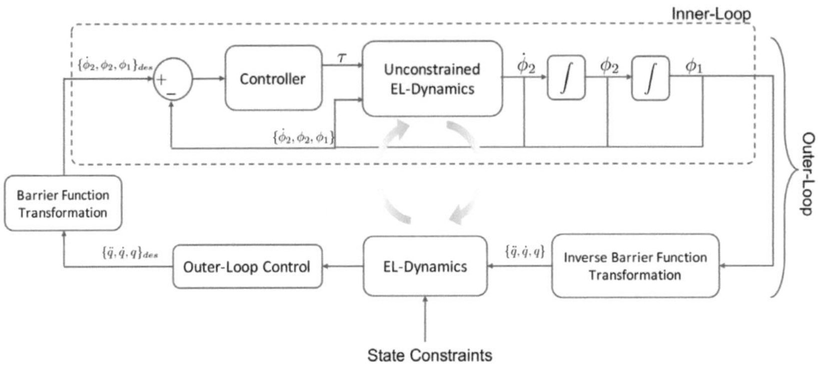

Figure 1.
A block diagram that illustrates the control flow in a robotic system. The architecture constitutes of two main loops. The inner-loop represents an equivalent unconstrained Euler–Lagrange (EL)-dynamics used to design an adaptive controller. The outer-loop contains the constrained EL-dynamics and a controller that defines the desired joint motions at each time step.

directions of the method and its applicability to safety in collaborative robotics are discussed.

Rest of the chapter is organized as follows. A review of BFs and CBFs and Barrier transformations is presented. Barrier transformation is then used to design adaptive robot controller of EL robot system in this chapter. A design and analysis of the safe adaptive trajectory tracking controller is then discussed. Simulation results of the designed controller on a 2-link EL robot system model are presented. Future directions of robot control design for safe human-robot collaboration are provided at the end.

2. Review of barrier functions and control barrier functions

In this section, a brief review of BF and CBF are presented.

2.1 Barrier functions

Consider a continuous nonlinear dynamical system of the form

$$\dot{x} = f(x), \tag{1}$$

where $f : \mathbb{R}^n \rightarrow \mathbb{R}^n$ is a locally Lipschitz continuous nonlinear function and $x(t) \in \mathcal{X} \subseteq \mathbb{R}^n$ is the state of the system. A set $\mathcal{S} \in \mathbb{R}^n$ is called *(forward) invariant* with respect to (1) if for any initial condition $x(0){:=}x(t_0) \in \mathcal{S}$ implies that $x(t) \in \mathcal{S}$, $\forall t \in \mathbb{R}$ [68]. BFs define a forward invariant safe region, where the solutions of a dynamical system in this region remain in the region for all time [46, 47, 69].

2.1.1 Constructing the barrier functions

Given a closed set $\mathcal{S} \subset \mathbb{R}^n$, its interior and its boundary are defined as follows

$$\mathcal{S} = \{x \in \mathbb{R}^n \ : \ h(x) \geq 0\}, \tag{2}$$

$$\partial\mathcal{S} = \{x \in \mathbb{R}^n \ : \ h(x) = 0\}, \tag{3}$$

$$Int(\mathcal{S}) = \{x \in \mathbb{R}^n \ : \ h(x) > 0\}, \tag{4}$$

where $h : \mathbb{R}^n \rightarrow \mathbb{R}$ is a continuously differentiable function.

Definition 1. *[47, Definition 1] Given the continuous system (1), the closed set $\mathcal{S}$ defined by (2)–(4), and continuously differentiable function $h : \mathbb{R}^n \rightarrow \mathbb{R}$, a real-valued function $\mathfrak{b} : Int(\mathcal{S}) \rightarrow \mathbb{R}$ that is differentiable with respect to its argument is said to be a reciprocal BF, if there exist class K functions η_1, η_2, η_3 such that for all $x \in Int(\mathcal{S})$*

$$\frac{1}{\eta_1(h(x))} \leq \mathfrak{b}(x) \leq \frac{1}{\eta_2(h(x))}, \tag{5}$$

$$\frac{\partial \mathfrak{b}(x)}{\partial x} f(x) \leq \eta_3(h(x)). \tag{6}$$

Candidate reciprocal BFs are *inverse-type* and *logarithmic-type* BFs given by $\mathfrak{b}(x) = \frac{1}{h(x)}$ and $\mathfrak{b}(x) = -\log \frac{h(x)}{1+h(x)}$, respectively [47]. Note that the candidate is unbounded on the set boundary, i.e., $\mathfrak{b}(x) \rightarrow \infty$ as $x \rightarrow \partial\mathcal{S}$.

2.2 Control barrier functions

BFs are essential means to verify invariance of a set but they cannot be used in its direct form to design a controller [47]. In other words, to make sure that the set $Int(\mathcal{S})$ is forward invariant under the dynamics of the system (1), a controller that guarantees the invariance of the set is required. Similar on how Lyapunov functions are extended to control Lyapunov functions [70], the concept of BFs can be extended to the case of control systems through the use of CBFs. Given the following nonlinear affine control system

$$\dot{x} = f(x) + g(x)u, \tag{7}$$

with f and g locally Lipschitz, $x \in \mathcal{X} \subset \mathbb{R}^n$, and $u \in \mathbb{R}^m$ is the set of admissible input, in cases where the solutions of (7) do not stay in an invariant set $\mathcal{S}$, a CBF can be specified that will assure the solutions to remain inside the invariant set.

2.2.1 Constructing the control barrier functions

In order to find a suitable CBF, the constraint on the system state x is encoded in a smooth constraint function $h(x)$. A value $h(x) > 0$ indicates adherence, whereas $h(x) < 0$ indicates a violation. The set of admissible state $\mathcal{X}_0$ is defined by

$$\mathcal{X}_0 = \{x \in \mathbb{R}^n : \ h(x) > 0\} \tag{8}$$

$$\partial\mathcal{X}_0 = \{x \in \mathbb{R}^n : \ h(x) = 0\} \tag{9}$$

A Reciprocal CBF $\mathfrak{b} : Int(\mathcal{X}_o) \rightarrow \mathbb{R}$ is a non-negative function, if there exist class K functions η_1,η_2, and η_3 such that for all $x \in Int(\mathcal{X}_0)$,

$$\frac{1}{\eta_1(h(x))} \leq \mathfrak{b}(x) \leq \frac{1}{\eta_2(h(x))} \tag{10}$$

$$\inf_{u \in \mathbb{R}^m} \{\mathcal{L}_f \mathfrak{b}(x) + \mathcal{L}_g \mathfrak{b}(x)u - \eta_2(h(x))\} \leq 0. \tag{11}$$

where $\mathcal{L}_f \mathfrak{b}(x)$ is the Lie-Derivative $\frac{\partial \mathfrak{b}(x)}{\partial x} f(x)$ along the vector field $f(x)$ and $\mathcal{L}_g \mathfrak{b}(x)$ is the Lie-Derivative $\frac{\partial \mathfrak{b}(x)}{\partial x} g(x)$ along the vector field $g(x)$. Hence for the system in (7), any locally Lipschitz controller $u : \mathcal{X}_0 \rightarrow \mathbb{R}^m$ that is selected form (11) assures the closed-set $\mathcal{X}_0 \subset \mathbb{R}^n$ is forward invariant.

3. Review of barrier transformation

In this section, review of barrier function transformation is presented. Consider the following logarithmic barrier function $B(z; a, A) : \mathbb{R} \to \mathbb{R}$ defined on an open interval (a, A):

$$B(z; a, A) \triangleq \ln\left(\frac{A}{a}\frac{a-x}{A-x}\right), \forall z \in (a, A). \tag{12}$$

where a and A are two constants satisfying $a < A$. The barrier function in (12) takes finite value when its arguments are within the region (a, A) and approaches to infinity as its arguments reach the boundary of the region, i.e., $\lim_{z \to a,A} B(z; a, A) = \pm\infty$.

Due to the monotonic characteristic of the natural logarithm the inverse of the barrier function (12) exists within the range of its definition, and it is given by

$$B^{-1}(y; a, A) = \frac{aA\left(e^{\frac{-y}{2}} - e^{\frac{y}{2}}\right)}{Ae^{\frac{-y}{2}} - ae^{\frac{y}{2}}}, \quad \forall y \in \mathbb{R} \tag{13}$$

with the derivative defined as

$$\frac{dB^{-1}(y; a, A)}{dy} = \frac{Aa^2 - aA^2}{a^2e^y - 2aA + A^2e^{-y}}. \tag{14}$$

4. Adaptive control of a robot system with full-state constraints

When a robot moves in a constrained space, it is crucial for the robot to satisfy requirements, such as the joint trajectories' boundedness, to safely carry out its operations within a prescribed bound. This section presents an adaptive safe tracking control design method that learns the parameters of an uncertain Euler–Lagrange (EL) system in an online manner using a gradient adaptive learning law. The controller is designed to track joint angles and joint velocities of the robot arm such that the bounds on the joint angles and joint velocities are maintained.

4.1 Euler-Lagrange dynamics for robot arm

Consider the Euler–Lagrange (EL) dynamics

$$M(q)\ddot{q} + C(q, \dot{q})\dot{q} + G_r(q) = \tau, \tag{15}$$

where $M(q) \in \mathbb{R}^{d \times d}$ denotes a generalized inertia matrix, $C(q, \dot{q}) \in \mathbb{R}^{d \times d}$ denotes a generalized centripetal-Coriolis matrix, $G_r(q) \in \mathbb{R}^d$ denotes a generalized gravity vector, $\tau = [\tau_1, \cdots, \tau_d]^T \in \mathbb{R}^d$ represents the generalized input control vector, and $q(t),\ \dot{q}(t),\ \ddot{q}(t) \in \mathbb{R}^d$ denote the link position, velocity, and acceleration vectors, respectively. The subsequent development is based on the assumption that all the states are observed, and that $M(q)$, $C(q, \dot{q})$, and $G_r(q)$, are unknown. The following properties, found in [71, 72], are also exploited in the subsequent development.

Property 1. The inertia matrix is positive definite, and satisfies the following inequality for any arbitrary vector $\xi \in \mathbb{R}^d$:

$$\mathbf{m}_1 \|\xi\|^2 \le \xi^T M(q) \xi \le \mathbf{m}_2 \|\xi\|^2, \tag{16}$$

where $\mathbf{m}_1$ and $\mathbf{m}_2$ are positive constants, and $\|\cdot\|$ represents the Euclidean norm.

Remark 1. Since $M(q)$ is a symmetric positive definite matrix, it can be shown that $M^{-1}(q)$ is also a positive definite matrix, and its 2-norm is upper and lower bounded with known constants, i.e., $\underline{m} \le \|M^{-1}(q)\| \le \overline{m}$.

Property 2. The EL-dynamics in (15) are linearly parametrizable as follows

$$Y(q,\dot{q},\ddot{q})\theta = M(q)\ddot{q} + C(q,\dot{q})\dot{q} + G_r(q), \tag{17}$$

where $Y: \mathbb{R}^d \times \mathbb{R}^d \times \mathbb{R}^d \to \mathbb{R}^{d\times m}$ is the regression matrix, and $\theta \in \mathbb{R}^n$ is the set of the unknown parameters.

Property 3. The norm of the centripetal-Coriolis can be upper bounded in the following manner:

$$\|C(q,\dot{q})\|_\infty \le \overline{C}\|\dot{q}\|, \tag{18}$$

where $\overline{C} \in \mathbb{R}$ denotes known positive bounding constant, and $\|\cdot\|_\infty$ denotes the induced infinity-norm of a matrix.

4.2 State space system model and control design

Let $x = [x_1, x_2]^T \in \mathcal{X} \subset \mathbb{R}^{2d}$, where $x_1 = q \in \mathbb{R}^d$, $x_2 = \dot{q} \in \mathbb{R}^d$, and the EL-dynamics in (15) can be written as follows

$$\begin{aligned} \dot{x}_1 &= x_2, \\ \dot{x}_2 &= \mathbf{f}(x) + \mathbf{g}(x)\tau, \end{aligned} \tag{19}$$

where $\mathbf{f}: \mathbb{R}^{2d} \to \mathbb{R}^d$, $\mathbf{g}: \mathbb{R}^{2d} \to \mathbb{R}^{d\times d}$ are locally Lipschitz continuous nonlinear functions, $\mathbf{f}(x) = M^{-1}(x_1)(-C(x_1,x_2)x_2 - G_r(x_1))$, and $\mathbf{g}(x) = M^{-1}(x_1)$. With some algebraic manipulations, the EL-dynamics can be written into d separate first and second order dynamics:

$$\dot{x}_{1,j} = x_{2,j}, \tag{20}$$

$$\dot{x}_{2,j} = f_j(x) + g_j(x)\tau, \quad \forall j = 1, \cdots, d \tag{21}$$

where $f_j: \mathbb{R}^{2d} \to \mathbb{R}$, $g_j: \mathbb{R}^{2d} \to \mathbb{R}^{1\times d}$ are nonlinear continuously differentiable functions. Using the BF transformation (12), the system in (20)–(21) can be transformed into a constrained state $\Phi = [\phi_1, \phi_2]^T \in \mathbb{R}^{2d}$, where $\phi_1 = [\varphi_{1,1}, \cdots, \varphi_{1,d}]^T$ and $\phi_2 = [\varphi_{2,1}, \cdots, \varphi_{2,d}]^T$ are the constrained joint position and velocity vectors, respectively, as follows:

$$\varphi_{i,j} = B(x_{i,j}; \delta_{i,j}, \Delta_{i,j}), \qquad \forall i = 1,2 \text{ and } \forall j = 1, \cdots, d \tag{22}$$

$$x_{i,j} = B^{-1}(\varphi_{i,j}; \delta_{i,j}, \Delta_{i,j}), \qquad \forall i = 1,2 \text{ and } \forall j = 1, \cdots, d \tag{23}$$

where $B^{-1}(\varphi_{i,j}; \delta_{i,j}, \Delta_{i,j})$ can be obtained using (13) and $\delta_{i,j}$, $\Delta_{i,j}$ are lower and upper bounds on state, respectively. Using the chain rule of differentiation, i.e., $\frac{dx_{i,j}}{dt} = \frac{\partial x_{i,j}}{\partial \varphi_{i,j}} \frac{d\varphi_{i,j}}{dt}$, where $\frac{\partial x_{i,j}}{\partial \varphi_{i,j}}$ can be obtained using (14), and some algebraic manipulations result in the transformed state $\varphi_{i,j}$, and it is given by

$$\dot{\varphi}_{1,j} = \mathcal{K}_{1,j}\left(\varphi_{1,j}\right)B^{-1}\left(\varphi_{2,j};\delta_{2,j},\Delta_{2,j}\right) = F_{1,j}\left(\varphi_{1,j},\varphi_{2,j}\right), \quad \forall j = 1,\cdots,d, \tag{24}$$

$$\dot{\varphi}_{2,j} = \mathcal{K}_{2,j}\left(\varphi_{2,j}\right)\left(f_j(x) + g_j(x)u\right) = F_{2,j}(\Phi) + G_{2,j}(\Phi)\tau, \quad \forall j = 1,\cdots,d, \tag{25}$$

where

$$F_{2,j}(\Phi) = \mathcal{K}_{2,j}\left(\varphi_{2,j}\right)f_j\left(\left[B^{-1}(\varphi_{1,1}) \quad \cdots \quad B^{-1}\left(\varphi_{2,j}\right)\right]\right) \tag{26}$$

$$G_{2,j}(\Phi) = \mathcal{K}_{2,j}\left(\varphi_{2,j}\right)g_j\left(\left[B^{-1}(\varphi_{1,1}) \quad \cdots \quad B^{-1}\left(\varphi_{2,j}\right)\right]\right) \tag{27}$$

and $\mathcal{K}_{i,j}\left(\varphi_{i,\varphi}\right) = \left(\frac{\partial x_{i,\varphi}}{\partial \varphi_{i,j}}\right)^{-1}$, $\forall i = 1,2$ and $\forall j = 1,\cdots,d$. The constrained system in terms of Φ can be expressed in a compact form as follows

$$\dot{\Phi} = \mathcal{F}(\Phi) + \mathcal{G}(\Phi)\tau, \tag{28}$$

where $\mathcal{F}: \mathbb{R}^{2d} \to \mathbb{R}^{2d}$ and $\mathcal{G}: \mathbb{R}^{2d} \to \mathbb{R}^{2d\times d}$ are given by

$$\mathcal{F}(\Phi) = \begin{bmatrix} F_{1,1}\left(\varphi_{1,1},\varphi_{2,1}\right) \\ \vdots \\ F_{1,d}\left(\varphi_{1,d},\varphi_{2,d}\right) \\ F_{2,1}(\Phi) \\ \vdots \\ F_{2,d}(\Phi) \end{bmatrix}, \quad \mathcal{G}(\Phi) = \begin{bmatrix} 0_{d\times d} \\ G_{2,1}(\Phi) \\ \vdots \\ G_{2,d}(\Phi) \end{bmatrix}. \tag{29}$$

Assumption 1. The function $\mathcal{F}: \mathbb{R}^{2d} \to \mathbb{R}^{2d}$ is locally Lipschitz continuous, and there exists a positive constant $\overline{\mathcal{F}}$ such that for $\Phi \in \Psi$, $\|\mathcal{F}(\Phi)\| < \overline{\mathcal{F}}\|\Phi\|$, where $\Psi \subset \mathbb{R}^{2d}$ is a compact set containing the origin. Moreover, the system is assumed to be controllable over Ψ with $\mathcal{G}(\Phi)$ being locally Lipschitz and bounded in Ψ, i.e., $\|\mathcal{G}(\Phi)\| < \overline{\mathcal{G}}$, where $\overline{\mathcal{G}}$ is a positive scalar.

Following (28), the EL-dynamics can be represented in the constrained space as follows

$$M\left(\varphi_p\right)\mathcal{K}_2^{-1}(\phi_2)\dot{\phi}_2 + C\left(\varphi_p,\varphi_v\right)\mathcal{K}_1^{-1}(\phi_1)\phi_2 + G_r\left(\varphi_p\right) = \tau, \tag{30}$$

where

$$\varphi_p = \left[B^{-1}(\varphi_{1,1}), \quad \cdots \quad, B^{-1}(\varphi_{1,d})\right]^T, \tag{31}$$

$$\varphi_v = \left[B^{-1}(\varphi_{2,1}), \quad \cdots \quad, B^{-1}(\varphi_{2,d})\right]^T, \tag{32}$$

and

$$\mathcal{K}_i(\phi_i) = \begin{bmatrix} \mathcal{K}_{i,1}\left(\varphi_{i,1}\right) & & 0 \\ & \ddots & \\ 0 & & \mathcal{K}_{i,d}\left(\varphi_{i,d}\right) \end{bmatrix}, \tag{33}$$

with $\mathcal{K}_{i,j}\left(\varphi_{i,j}\right) = \frac{\partial B^{-1}\left(\varphi_{i,j}\right)}{\partial \varphi_{i,j}}$, $\forall i = 1,2$ and $\forall j = 1,\cdots,d$.

Assumption 2. The terms $\mathcal{K}_i(\phi_i)$ defined in (33) is positive definite, and its 2-norm is upper and lower bounded by known positive constants, i.e., $\underline{k}_i \leq \|\mathcal{K}_i(\phi_i)\| \leq \overline{k}_i, \forall i = 1, 2$.

Lemma 1. Given the term $\mathcal{K}_i(\phi_i)$ defined in (33) with

$$\mathcal{K}_{i,j}\left(\varphi_{i,j}\right) = \frac{\left(\delta_{i,j}^2 e^{\varphi_{i,j}} - 2\delta_{i,j}\Delta_{i,j} + \Delta_{i,j}^2 e^{-\varphi_{i,j}}\right)}{\Delta_{i,j}\delta_{i,j}^2 - \delta_{i,j}\Delta_{i,j}^2}, \quad \forall i = 1, 2 \text{ and } \forall j = 1, \cdots, d, \tag{34}$$

the 2-norm of its inverse, $\mathcal{K}_i^{-1}(\phi_i)$, can be upper bounded by a positive constant $\overline{\kappa}_i$, i.e., $\|\mathcal{K}_i^{-1}(\phi_i)\| \leq \overline{\kappa}_i, \forall i = 1, 2$.

Proof: The 2-norm of $\mathcal{K}_i^{-1}(\phi_i) = \text{diag}\left(\mathcal{K}_{i,1}^{-1}(\varphi_{i,1}), \cdots, \mathcal{K}_{i,d}^{-1}(\varphi_{i,d})\right)$ can be upper bounded because $\mathcal{K}_{i,j}^{-1}\left(\varphi_{i,j}\right)$ is bounded, that is

$$\lim_{\varphi_{i,j} \to \infty} \frac{\Delta_{i,j}\delta_{i,j}^2 - \delta_{i,j}\Delta_{i,j}^2}{\left(\delta_{i,j}^2 e^{\varphi_{i,j}} - 2\delta_{i,j}\Delta_{i,j} + \Delta_{i,j}^2 e^{-\varphi_{i,j}}\right)} = 0, \tag{35}$$

which implies that 2-norm of $\mathcal{K}_i^{-1}(\varphi_i)$ can be upper bounded by a positive constant $\overline{\kappa}_i$.

Now, using Property 2, the EL-dynamics in (30) can be linearly parameterized, and it is given by

$$M\mathcal{K}_2^{-1}\dot{\phi}_2 + C\mathcal{K}_1^{-1}\phi_2 + G_r = Y_1\left(\varphi_p, \varphi_v, \phi_1, \phi_2, \dot{\phi}_2\right)\theta, \tag{36}$$

where $Y_1 : \mathbb{R}^d \times \mathbb{R}^d \times \mathbb{R}^d \times \mathbb{R}^d \times \mathbb{R}^d \to \mathbb{R}^{d \times n}$ is the regression matrix. Note that in (36), and henceforth the parameter dependency of the elements in the EL-dynamics are dropped for brevity.

Lemma 2. Suppose that there exists a controller that tracks the desired trajectory for the system given in (30). Then, the same controller can also track the desired trajectory of the original system in (15) given that the initial state of the system $x(0) = x_0 \in \mathcal{X}$.

Proof: See proof of ([55], Lemma 1)

Lemma 2 proves that if the initial state is within the prescribed bound, a control law can be designed for the full-state constrained system such that it satisfies the tracking objective of the original system.

4.2.1 Safe adaptive tracking control development

In this subsection, an adaptive control technique is used to identify the parameters of an uncertain system and track the desired joint position $\phi_1^{des}(t) : \mathbb{R}^+ \to \mathbb{R}^d$ and joint velocity $\phi_2^{des}(t) : \mathbb{R}^+ \to \mathbb{R}^d$ trajectories.

Assumption 3. The signals $\phi_1^{des}, \phi_2^{des}, \dot{\phi}_2^{des}$ are uniformly continuous and bounded such that $\|\phi_1^{des}\| \leq \overline{\phi}_1^{des}$, $\|\phi_2^{des}\| \leq \overline{\phi}_2^{des}$, $\|\dot{\phi}_2^{des}\| \leq \overline{\dot{\phi}}_2^{des}$, where $\overline{\phi}_1^{des}$, $\overline{\phi}_2^{des}$, and $\overline{\dot{\phi}}_2^{des}$ are known positive constants.

Consider the following tracking control input design

$$\tau = \hat{M}\mathcal{K}_2^{-1}a + \hat{C}\mathcal{K}_1^{-1}v + \hat{G}_r - \beta\mathcal{K}_2 r, \tag{37}$$

where $(\hat{\cdot})$ denotes the parameter estimates and β is a positive scalar. Signals a, v, r are given by

$$a = \dot{\phi}_2^{des} - \Lambda\tilde{\phi}_2, \tag{38}$$

$$v = \phi_2^{des} - \Lambda\tilde{\phi}_1, \tag{39}$$

$$r = \tilde{\phi}_2 + \Lambda\tilde{\phi}_1, \tag{40}$$

where $\tilde{\phi}_1 \triangleq \phi_1 - \phi_1^{des}$ and $\tilde{\phi}_2 \triangleq \phi_2 - \phi_2^{des}$ are position and velocity tracking errors, respectively. $\Lambda \in \mathbb{R}^{d\times d}$ is a positive definite diagonal matrix, and its 2-norm is upper bounded by a known positive constant, i.e., $\|\Lambda\| \leq \overline{\Lambda}$.

In terms of the linear parameterization of the EL-dynamics, i.e., Property 2, the control input (37) can be rewritten as

$$\tau = Y_2\left(\varphi_p, \varphi_v, \mathcal{K}_2^{-1}(\phi_2)a, \mathcal{K}_1^{-1}(\phi_1)v\right)\hat{\theta} - \beta\mathcal{K}_2(\phi_2)r, \tag{41}$$

where $Y_2 : \mathbb{R}^d \times \mathbb{R}^d \times \mathbb{R}^d \times \mathbb{R}^d \to \mathbb{R}^{d\times n}$ is the regression matrix. Substituting (37) in the EL-dynamics (30) yields the following closed-loop error dynamics given by

$$M\mathcal{K}_2^{-1}\dot{r} + C\mathcal{K}_1^{-1}r + \beta\mathcal{K}_2 r = Y_2\tilde{\theta}, \tag{42}$$

where $\tilde{\theta} = \hat{\theta} - \theta$ is the parameter estimation error. The parameter $\hat{\theta}$ update rule is given by

$$\dot{\hat{\theta}} = \text{proj}\left(-\Gamma^{-1}Y_2^T\mathcal{K}_2 r\right), \tag{43}$$

where $\Gamma \in \mathbb{R}^{n\times n}$ is a diagonal and positive definite matrix, and proj$(\cdot)$ is a standard projection operator that ensures the parameter estimates are bounded, i.e., $\underline{\theta} \leq \hat{\theta} \leq \overline{\theta}$ (for further details see [71]).

Remark 2. The parameter estimation error $\tilde{\theta}$ is bounded and uniformly continuous since $\hat{\theta}$ evolves according to the update law in (43).

4.2.2 Lyapunov stability analysis

To facilitate the following development of the Lyapunov stability analysis, let $\zeta : [0,\infty) \to \mathbb{R}^{2d+n}$ denote the composite state vector, i.e., $\zeta(t) \triangleq \left[r^T(t), \phi_1^T(t), \tilde{\theta}^T(t)\right]^T$. Let $\lambda_{min}\{\cdot\}$ and $\lambda_{max}\{\cdot\}$ denote the minimum and maximum eigenvalues of its argument.

Theorem 1.1. *The controller and parameter update laws defined in (41) and (43) ensure SGUUB tracking of the desired state trajectories, provided the following sufficient conditions,*

$$\gamma_1 > 2(1+\gamma_5+\gamma_7), \qquad \gamma_3 > 2(1+\gamma_5+\gamma_6), \tag{44}$$

are satisfied, where

$$\begin{array}{ll} \gamma_1 = \beta\underline{m} & \underline{k}_2^2\gamma_5 = \beta\overline{\lambda}\overline{k}_2^2\overline{m} \\ \gamma_2 = \lambda_{max}\{\Lambda^T\Lambda\}\overline{\alpha} & \gamma_6 = (\gamma_2+\gamma_4)\overline{\phi}_2^{des} \\ \gamma_3 = \gamma_1\lambda_{min}\{\Lambda^T\Lambda\} & \gamma_7 = (\overline{\alpha}+\gamma_4)\overline{\phi}_2^{des} \\ \gamma_4 = \overline{\Lambda}\overline{\alpha} & \overline{\alpha} = \overline{k}_2\overline{m}\overline{C}\overline{\kappa}_1^2 \end{array} \tag{45}$$

Proof: See details in [67].

5. Simulations

Simulation studies are conducted to verify and demonstrate the performance of the designed safe adaptive robot controller. The simulations are conducted using MacBook Pro running Intel i7 processor and 16 Gigabytes of memory and the controller and EL dynamic model is coded using MATLAB 2018a.

5.1 Safe tracking control of an uncertain EL-dynamics with full-state constraints using BF

In this section, the controller and adaptive laws developed in (37) and (43) are simulated for a two-link robot planar manipulator, with dynamics shown in (46), where c_1, c_2, c_{12} denote $\cos(q_1)$, $\cos(q_2)$, and $\cos(q_1+q_2)$ respectively, $\sin_2$ denotes $\sin(q_2)$, and g is the gravitational constant.

$$\underbrace{\begin{bmatrix} \theta_1+2\theta_2 c_2 & \theta_3+\theta_2 c_2 \\ \theta_3+\theta_2 c_2 & \theta_3 \end{bmatrix}}_{M(q)} \begin{bmatrix} \ddot{q}_1 \\ \ddot{q}_2 \end{bmatrix} + \underbrace{\begin{bmatrix} -\theta_2 \sin_2 \dot{q}_2 & -\theta_2 \sin_2 (\dot{q}_1+\dot{q}_2) \\ \theta_2 \sin_2 \dot{q}_1 & 0 \end{bmatrix}}_{C(q,\dot{q})} \begin{bmatrix} \dot{q}_1 \\ \dot{q}_2 \end{bmatrix} + \underbrace{\begin{bmatrix} \theta_4 g c_1 + \theta_5 g c_{12} \\ \theta_5 g c_{12} \end{bmatrix}}_{G_r(q)} = \begin{bmatrix} \tau_1 \\ \tau_2 \end{bmatrix} \tag{46}$$

The nominal values of the parameter vector $\theta = [\theta_1, \theta_2, \theta_3, \theta_4, \theta_5]^T$ are

$$\begin{array}{lll} \theta_1 = 0.325 \ \text{kg} \cdot \text{m}^2 & \theta_3 = 0.217 \ \text{kg} \cdot \text{m}^2 & \\ \theta_2 = 0.240 \ \text{kg} \cdot \text{m}^2 & \theta_4 = 2.4 \ \text{kg} \cdot \text{m} & \theta_5 = 1.0 \ \text{kg} \cdot \text{m} \end{array} \tag{47}$$

The desired trajectory is selected as

$$q_{d_1} = (-4 - 6e^{-2t}) \sin(t), q_{d_2} = (-4 - 3e^{-t}) \cos(t). \tag{48}$$

The objective is to track the desired joint trajectory provided that the model parameters are unknown while the state $Q = [q, \ \dot{q}]^T$ satisfies the following constraints,

$$\begin{array}{ll} q_1 \in (-4.4, 4.1) & \dot{q}_1 \in (-10.2, 4.2) \\ q_2 \in (-7.1, 4.2) & \dot{q}_2 \in (-4.2, 4.93) \end{array} \tag{49}$$

To this end, the barrier function formulation presented in Section 3 is used along with the adaptive control developed in Section 4. The feedback and adaptation gains for the proposed controller are selected as $\beta = 14$, $\Lambda = \text{diag}(2.01, 2.01)$, and $\Gamma = \text{diag}(30, 30)$. The results of the simulation are shown in **Figures 2–4**. The joints

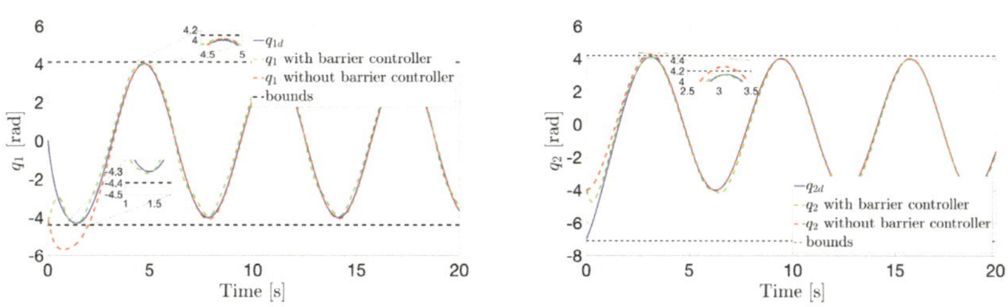

Figure 2.
Evolution of the joint angles for the planar robot simulation using an adaptive law with and without BF.

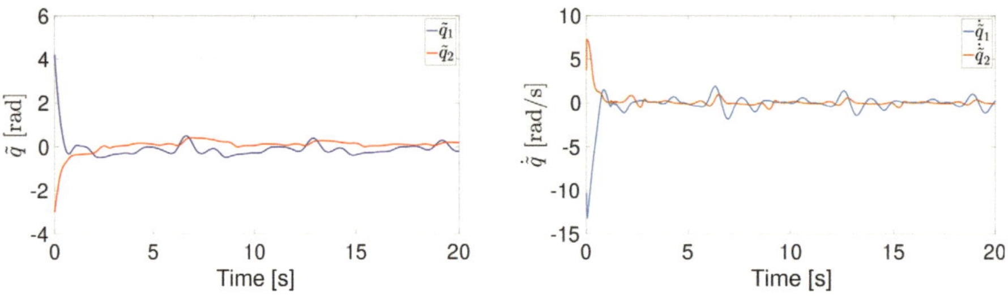

Figure 3.
Evolution of the joint angle errors and joint velocity errors for the planar robot simulation using an adaptive law with BF.

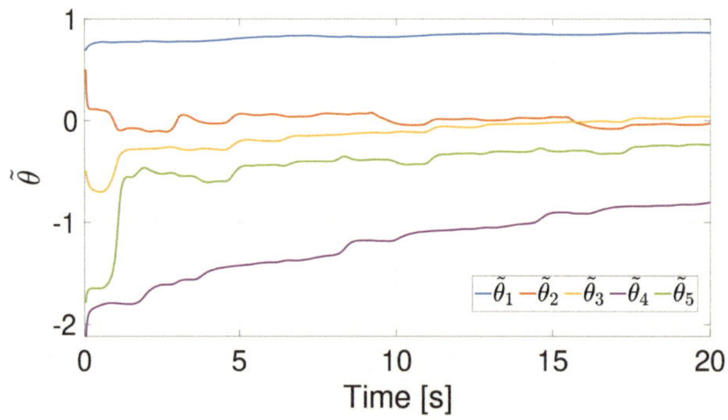

Figure 4.
Evolution of the parameter estimation error for the planar robot simulation.

position evolution $q_1(t)$ and $q_2(t)$ of a two degrees-of-freedom planar robot using an adaptive law with and without BF are shown in **Figure 2**. It can be observed from **Figure 2** that when the adaptive law with BF is used, the estimated trajectories are blocked from crossing over the boundaries that are set for each of the joints. The position and velocity estimation errors are depicted in **Figure 3**. From **Figures 2** and **3**, it is clear that the tracking error asymptotically converges to zero, and, because the Lyapunov candidate does not contain any terms that are negative definite in $\tilde{\theta}$, the parameter estimation does not converge but it does remain bounded. Boundedness of the parameter estimation errors can be seen in **Figure 4**.

6. Conclusions and future directions

This chapter provides a perspective on problems wherein humans and robots work collaboratively with one another. Research in this field aims to relax the current workplace constraints, such as fences, virtual curtains often seen in manufacturing settings between humans and robots or velocity limits on collaborative robots. This chapter develops an efficient robot control methodology to create a safe working environment without sacrificing the efficiency of the robots. In the context of the chapter, safety is defined as a constrained behavior of a system, and robot effectiveness, as driving the actual behavior of the robot to the desired behavior. To this end, an online safe tracking controller for an uncertain Euler–Lagrange robotic system with is developed where the constraints are placed on all the states. A barrier function transform is used to transform the full-state constrained EL-dynamics into an equivalent unconstrained system with no prior knowledge of the system parameters. An adaptive controller is developed along with a gradient based adaptive parameter estimation law on the transformed system that tracks the desired trajectories of the original system. The controller guarantees that the robot trajectories remain inside a pre-specified safe region, tracking the desired trajectories and the parameter estimation errors remain bounded. The method can be utilized for applications wherein robots must operate in a confined space to reach an object for grasping or other manipulation tasks such as pick and place.

In future, the usefulness of barrier transformation to design a visual servo controller will be shown. Constrained VS approach can guarantee target features to remain within the camera field of view for the duration of the task. Some recent efforts in that direction can be found in [73]. Utilizing CBF for developing safe robot controllers by utilizing human actions and workspaces can be another avenue of future research for safe human-robot interaction.

Acknowledgements

The authors would like to thank Daniel Trombetta for discussions related to human-robot interaction application and control design.

Author details

Iman Salehi[1], Ghananeel Rotithor[2] and Ashwin Dani[1*]

1 University of Connecticut, Storrs, USA

2 Electrical and Computer Engineering, University of Connecticut, Storrs, CT, USA

*Address all correspondence to: ashwin.dani@uconn.edu

References

[1] S. Haddadin, Towards safe robots: approaching Asimov's 1st law. Springer, 2013, vol. 90.

[2] A. De Luca, A. Albu-Schaffer, S. Haddadin, and G. Hirzinger, "Collision detection and safe reaction with the dlr-iii lightweight manipulator arm," in *2006 IEEE/RSJ International Conference on Intelligent Robots and Systems,* 2006, pp. 1623–1630.

[3] F. Flacco and A. De Luca, "Safe physical human-robot collaboration," in *IEEE/RSJ International Conference on Intelligent Robots and Systems,* 2013, pp. 2072–2072.

[4] M. A. Peshkin, J. E. Colgate, W. Wannasuphoprasit, C. A. Moore, R. B. Gillespie, and P. Akella, "Cobot architecture," IEEE Transactions on Robotics and Automation, vol. 17, no. 4, pp. 377–390, 2001.

[5] P. A. Lasota, T. Fong, J. A. Shah *et al., A survey of methods for safe human-robot interaction.* Now Publishers, 2017.

[6] Y.-C. Peng, D. S. Carabis, and J. T. Wen, "Collaborative manipulation with multiple dual-arm robots under human guidance," International Journal of Intelligent Robotics and Applications, vol. 2, no. 2, pp. 252–266, 2018.

[7] H. Maske, M. Matthews, A. Axelrod, H. Mohamadipanah, G. Chowdhary, C. Crick, and P. Pagilla, "Collaborative goal and policy learning from human operators of construction co-robots," in *Neural Information Processing Systems (NIPS),* 2014.

[8] J. Braasch, R. J. Radke, J. Wen, M. Si, A. Cunningham, W. Keddy-Hector, and U. Sinha, "Audio/visual concepts for human/robot communication in immersive virtual environments," The Journal of the Acoustical Society of America, vol. 135, no. 4, pp. 2169–2169, 2014.

[9] T. Fong, C. Thorpe, and C. Baur, "Multi-robot remote driving with collaborative control," IEEE Transactions on Industrial Electronics, vol. 50, no. 4, pp. 699–704, 2003.

[10] T. Yucelen, Y. Yildiz, R. Sipahi, E. Yousefi, and N. Nguyen, "Stability limit of human-in-the-loop model reference adaptive control architectures," International Journal of Control, vol. 91, no. 10, pp. 2314–2331, 2018.

[11] D. Panagou, D. M. Stipanović, and P. G. Voulgaris, "Distributed coordination control for multi-robot networks using lyapunov-like barrier functions," IEEE Transactions on Automatic Control, vol. 61, no. 3, pp. 617–632, 2015.

[12] Z. Kan, A. P. Dani, J. M. Shea, and W. E. Dixon, "Network connectivity preserving formation stabilization and obstacle avoidance via a decentralized controller," IEEE Transactions on Automatic Control, vol. 57, no. 7, pp. 1827–1832, 2012.

[13] A. W. Farras, J. Yamauchi, T. Hatanaka, and M. Fujita, "Safe cooperative control of human-robotic network teaming with control barrier function," in *2020 SICE International Symposium on Control Systems (SICE ISCS)*, 2020, pp. 33–39.

[14] K. Hawkins, N. Vo, S. Bansal, and A. Bobick, "Probabilistic human action prediction and wait-sensitive planning for responsive humanrobot collaboration," in *Proceedings of the IEEE-RAS International Conference on Humanoid Robots*, 2013, pp. 15–17.

[15] P. A. Lasota, G. F. Rossano, and J. A. Shah, "Toward safe close-proximity human-robot interaction with standard industrial robots," in *IEEE International Conference on Automation Science and Engineering (CASE),* 2014, pp. 339–344.

[16] B. Alqaudi, H. Modares, I. Ranatunga, S. M. Tousif, F. L. Lewis, and D. O. Popa, "Model reference adaptive impedance control for physical human-robot interaction," Control Theory and Technology, vol. 14, no. 1, pp. 68–82, 2016.

[17] C.-S. Tsai, J.-S. Hu, and M. Tomizuka, "Ensuring safety in human-robot coexistence environment," in *IEEE/RSJ International Conference on Intelligent Robots and Systems (IROS)*, 2014, pp. 4191–4196.

[18] D.-J. Kim, Z. Wang, N. Paperno, and A. Behal, "System design and implementation of UCF-MANUS – an intelligent assistive robotic manipulator," IEEE/ASME transactions on mechatronics, vol. 19, no. 1, pp. 225–237, 2012.

[19] F. Zhang and H. Huang, "Source selection for real-time user intent recognition toward volitional control of artificial legs," IEEE journal of biomedical and health informatics, vol. 17, no. 5, pp. 907–914, 2012.

[20] D. De Carli, E. Hohert, C. A. Parker, S. Zoghbi, S. Leonard, E. Croft, and A. Bicchi, "Measuring intent in human-robot cooperative manipulation," in *IEEE International Workshop on Haptic Audio visual Environments and Games*. IEEE, 2009, pp. 159–163.

[21] Y. Li and S. Ge, "Human-robot collaboration based on motion intention estimation," IEEE/ASME Transactions on Mechatronics, vol. 19, no. 3, pp. 1007–1014, 2014.

[22] C. Yang, G. Ganesh, S. Haddadin, S. Parusel, A. Albu-Schaeffer, and E. Burdet, "Human-like adaptation of force and impedance in stable and unstable interactions," IEEE transactions on robotics, vol. 27, no. 5, pp. 918–930, 2011.

[23] N. Hogan, "Impedance control: An approach to manipulation: Part I—Theory, part II—Implementation, part III—Applications," Trans. ASME J. Dyn. Syst., Meas. Control, vol. 107, no. 1, pp. 1–24, 1985.

[24] I. Ranatunga, F. L. Lewis, D. O. Popa, and S. M. Tousif, "Adaptive admittance control for human–robot interaction using model reference design and adaptive inverse filtering," IEEE Transactions on Control Systems Technology, vol. 25, no. 1, pp. 278–285, 2016.

[25] K. Chen, Y. Zhang, J. Yi, and T. Liu, "An integrated physical-learning model of physical human-robot interactions with application to pose estimation in bikebot riding," The International Journal of Robotics Research, vol. 35, no. 12, pp. 1459–1476, 2016.

[26] S. Calinon, I. Sardellitti, and D. G. Caldwell, "Learning-based control strategy for safe human-robot interaction exploiting task and robot redundancies," in *IEEE/RSJ International Conference on Intelligent Robots and Systems*. IEEE, 2010, pp. 249–254.

[27] G. Hoffman and C. Breazeal, "Effects of anticipatory perceptual simulation on practiced human-robot tasks," Autonomous Robots, vol. 28, no. 4, pp. 403–423, 2010.

[28] S. Ferguson, B. Luders, R. C. Grande, and J. P. How, "Real-time predictive modeling and robust avoidance of pedestrians with uncertain, changing intentions," in *Algorithmic Foundations of Robotics XI*. Springer, 2015, pp. 161–177.

[29] F. Flacco, T. Kroger, A. De Luca, and O. Khatib, "A depth space approach to human-robot collision avoidance," in *IEEE International Conference on Robotics and Automation*, 2012, pp.338–345.

[30] M. Geravand, F. Flacco, and A. De Luca, "Human-robot physical

interaction and collaboration using an industrial robot with a closed control architecture," in *IEEE International Conference on Robotics and Automation*, 2013, pp. 4000–4007.

[31] O. Oshin, E. A. Bernal, B. M. Nair, J. Ding, R. Varma, R. W. Osborne, E. Tunstel, and F. Stramandinoli, "Coupling deep discriminative and generative models for reactive robot planning in human-robot collaboration," in *2019 IEEE International Conference on Systems, Man and Cybernetics (SMC)*, 2019, pp. 1869–1874.

[32] H. Ravichandar and A. P. Dani, "Human intention inference using expectation-maximization algorithm with online model learning," IEEE Transactions on Automation Science and Engineering, vol. 14, no. 2, pp. 855–868, 2017.

[33] H. C. Ravichandar, A. Kumar, and A. P. Dani, "Gaze and motion information fusion for human intention inference," International Journal of Intelligent Robotics and Applications, vol. 2, no. 2, pp. 136–148, 2018.

[34] A. P. Dani, I. Salehi, G. Rotithor, D. Trombetta, and H. Ravichandar, "Human-in-the-loop robot control for human-robot collaboration: Human intention estimation and safe trajectory tracking control for collaborative tasks," IEEE Control Systems Magazine, vol. 40, no. 6, pp. 29–56, 2020.

[35] C. Liu, J. B. Hamrick, J. F. Fisac, A. D. Dragan, J. K. Hedrick, S. S. Sastry, and T. L. Griffiths, "Goal inference improves objective and perceived performance in human-robot collaboration," in *International Conference on Autonomous Agents & Multiagent Systems*. International Foundation for Autonomous Agents and Multiagent Systems, 2016, pp. 940–948.

[36] J. Mainprice and D. Berenson, "Human-robot collaborative manipulation planning using early prediction of human motion," in *Intelligent Robots and Systems, 2013 IEEE/RSJ International Conference on*. IEEE, 2013, pp. 299–306.

[37] R. Wilcox, S. Nikolaidis, and J. Shah, "Optimization of temporal dynamics for adaptive human-robot interaction in assembly manufacturing," Robotics, pp. 441–449, 2013.

[38] V. Shivashankar, K. N. Kaipa, D. S. Nau, and S. K. Gupta, "Towards integrating hierarchical goal networks and motion planners to support planning for human-robot teams," in *AAAI Fall Symposium: Artificial Intelligence for Human-Robot Interaction*, 2014, pp. 13–15.

[39] E. Yousefi, Y. Yildiz, R. Sipahi, and T. Yucelen, "Stability analysis of a human-in-the-loop telerobotics system with two independent time-delays," IFAC-PapersOnLine, vol. 50, no. 1, pp. 6519–6524, 2017.

[40] I. Kolmanovsky, E. Garone, and S. Di Cairano, "Reference and command governors: A tutorial on their theory and automotive applications," in *American Control Conference*. IEEE, 2014, pp. 226–241.

[41] U. V. Kalabić, I. V. Kolmanovsky, and E. G. Gilbert, "Reduced order extended command governor," Automatica, vol. 50, no. 5, pp. 1466–1472, 2014.

[42] F. Flacco, A. De Luca, and O. Khatib, "Control of redundant robots under hard joint constraints: Saturation in the null space," IEEE Transactions on Robotics, vol. 31, no. 3, pp. 637–654, 2015.

[43] N. Fischer, A. Dani, N. Sharma, and W. E. Dixon, "Saturated control of an uncertain nonlinear system with input delay," Automatica, vol. 49, no. 6, pp. 1741–1747, 2013.

[44] F. Blanchini, "Set invariance in control," Automatica, vol. 35, no. 11, pp. 1747–1767, 1999.

[45] S. Prajna and A. Jadbabaie, "Safety verification of hybrid systems using barrier certificates," in *International Workshop on Hybrid Systems: Computation and Control,* 2004, pp. 477–492.

[46] X. Xu, P. Tabuada, J. W. Grizzle, and A. D. Ames, "Robustness of control barrier functions for safety critical control," IFAC-PapersOnLine, vol. 48, no. 27, pp. 54–61, 2015.

[47] A. D. Ames, X. Xu, J. W. Grizzle, and P. Tabuada, "Control barrier function based quadratic programs for safety critical systems," IEEE Transactions on Automatic Control, vol. 62, no. 8, pp. 3861–3876, 2017.

[48] F. Berkenkamp, R. Moriconi, A. P. Schoellig, and A. Krause, "Safe learning of regions of attraction for uncertain, nonlinear systems with gaussian processes," in *IEEE Conference on Decision and Control*, 2016, pp. 4661–4666.

[49] L. Wang, E. A. Theodorou, and M. Egerstedt, "Safe learning of quadrotor dynamics using barrier certificates," in *IEEE International Conference on Robotics and Automation*. IEEE, 2018, pp. 2460–2465.

[50] Q. Nguyen, A. Hereid, J. W. Grizzle, A. D. Ames, and K. Sreenath, "3d dynamic walking on stepping stones with control barrier functions," in *Decision and Control (CDC), 2016 IEEE 55th Conference on.* IEEE, 2016, pp. 827–834.

[51] G. Wu and K. Sreenath, "Safety-critical control of a planar quadrotor," in *American Control Conference (ACC), 2016*. IEEE, 2016, pp. 2252–2258.

[52] Q. Nguyen and K. Sreenath, "Exponential control barrier functions for enforcing high relative-degree safety-critical constraints," in *2016 American Control Conference (ACC)*, 2016, pp. 322–328.

[53] Q. Nguyen and K. Sreenath, "Safety-critical control for dynamical bipedal walking with precise footstep placement," IFAC-PapersOnLine, vol. 48, no. 27, pp. 147–154, 2015.

[54] Y. Yang, Y. Yin, W. He, K. G. Vamvoudakis, H. Modares, and D. C. Wunsch, "Safety-aware reinforcement learning framework with an actor-critic-barrier structure," in *American Control Conference*, 2019, pp. 2352–2358.

[55] K. P. Tee, S. S. Ge, and E. H. Tay, "Barrier lyapunov functions for the control of output-constrained nonlinear systems," Automatica, vol. 45, no. 4, pp. 918–927, 2009.

[56] K. B. Ngo, R. Mahony, and Z.-P. Jiang, "Integrator backstepping using barrier functions for systems with multiple state constraints," in *IEEE Conference on Decision and Control*, 2005, pp. 8306–8312.

[57] K. P. Tee, B. Ren, and S. S. Ge, "Control of nonlinear systems with time-varying output constraints," Automatica, vol. 47, no. 11, pp. 2511–2516, 2011.

[58] K. P. Tee and S. S. Ge, "Control of nonlinear systems with partial state constraints using a barrier lyapunov function," International Journal of Control, vol. 84, no. 12, pp. 2008–2023, 2011.

[59] W. He, Y. Chen, and Z. Yin, "Adaptive neural network control of an uncertain robot with full-state constraints," IEEE Transactions on Cybernetics, vol. 46, no. 3, pp. 620–629, March 2016.

[60] M. L. Greene, P. Deptula, S. Nivison, and W. E. Dixon, "Sparse learning-based approximate dynamic

programming with barrier constraints," IEEE Control Systems Letters, pp. 1–1, 2020.

[61] G. P. Kontoudis and K. G. Vamvoudakis, "Kinodynamic motion planning with continuous-time q-learning: An online, model-free, and safe navigation framework," IEEE Transactions on Neural Networks and Learning Systems, vol. 30, no. 12, pp. 3803–3817, 2019.

[62] J. F. Fisac, A. K. Akametalu, M. N. Zeilinger, S. Kaynama, J. Gillula, and C. J. Tomlin, "A general safety framework for learning-based control in uncertain robotic systems," IEEE Transactions on Automatic Control, vol. 64, no. 7, pp. 2737–2752, 2018.

[63] T. Schouwenaars, J. How, and E. Feron, "Receding horizon path planning with implicit safety guarantees," in *Proceedings of the American Control Conference*, 2004, pp. 5576–5581.

[64] L. S. Breger and J. P. How, "Safe trajectories for autonomous rendezvous of spacecraft," Journal of Guidance, Control, and Dynamics, vol. 31, no. 5, pp. 1478–1489, 2008.

[65] Y. Mao, D. Dueri, M. Szmuk, and B. Açıkmeşe, "Successive 20 convexification of non-convex optimal control problems with state constraints," in *IFAC-PapersOnLine*, 2017, pp. 4063–4069.

[66] N. Hovakimyan, C. Cao, E. Kharisov, E. Xargay, and I. M. Gregory, "l1 adaptive control for safety-critical systems," IEEE Control Systems Magazine, vol. 31, no. 5, pp. 54–104, 2011.

[67] I. Salehi, G. Rotithor, D. Trombetta, and A. P. Dani, "Safe tracking control of an uncertain euler-lagrange system with full-state constraints using barrier functions," in *59th IEEE Conference on Decision and Control (CDC)*. IEEE, 2020, pp. 3310–3315.

[68] H. K. Khalil, *Nonlinear Systems*, 3rd ed. Prentice Hall, 2002.

[69] S. Prajna, A. Jadbabaie, and G. J. Pappas, "Stochastic safety verification using barrier certificates," in *IEEE Conference on Decision and Control*, vol. 1. IEEE, 2004, pp. 929–934.

[70] E. D. Sontag, "A lyapunov-like characterization of asymptotic controllability," SIAM Journal on Control and Optimization, vol. 21, no. 3, pp. 462–471, 1983.

[71] W. E. Dixon, "Adaptive regulation of amplitude limited robot manipulators with uncertain kinematics and dynamics," IEEE Transactions on Automatic Control, vol. 52, no. 3, pp. 488–493, 2007.

[72] M. W. Spong, S. Hutchinson, and M. Vidyasagar, *Robot modeling and control*. Wiley New York, 2006, vol. 3.

[73] R. Funada, M. Santos, J. Yamauchi, T. Hatanaka, M. Fujita, and M. Egerstedt, "Visual coverage control for teams of quadcopters via control barrier functions," in *2019 International Conference on Robotics and Automation (ICRA)*, 2019, pp. 3010–3016.

Chapter 9

Tiny Blind Assistive Humanoid Robot

Amrita Ganguly and Bijan Paul

Abstract

In today's world, individuals show more enthusiasm for robotics and aim to depend upon humanoid robots for multiple purposes. It has implementations in a wide range of sectors such as- in atomic plants, house management, government foundations and even astro stations. Our research-based project is elucidating robotics with a tiny humanoid robot following the human structure to make gestures like strolling, dancing, and detecting objects near it. It has been achieved using Arduino Nano (Atmega 328P), Servo motors SG90 working on the conception of servo-mechanism and the Ultrasonic sensor to identify obstacles and restrain the Robot from going ahead. The aim is to fabricate a bigger humanoid robot that will serve our general public and make our life simpler. The Robot has eventual utilize in marketing, entertainment and helping the visually impaired to move from one place to another. This project can also be a great apparatus for future research works and alteration.

Keywords: bipedal Robot, Servo Mechanism, Object Detection, Ultrasonic sensor, Servo motor, Blind Assistive/Visually Impaired, entertainment, humanoid robot, notifier, rotation impact, navigation

1. Introduction

Individuals are finding robots charming and interesting day by day. In this changing climate humanoid robots have a more imperative capacity to outlive. One can without much of a stretch expect that in the coming future the utilization of robots will extend in our general public. It is only our little activity towards the development of a tiny humanoid to make the copy of human in the future. **Figure 1** shows the structure of the tiny humanoid strolling and moving Robot.

This Arduino-based bipedal Robot (**Figure 1**) follows an algorithm to stroll by turning the servo horns [1] and an ultrasonic sensor that can distinguish any items close to it [2]. At whatever point the humanoid identifies any item before it, it quits strolling and illuminates the LED on its top as an informer. This informer can be supplanted by any kind of informer like a buzzer [3]. A microcontroller ATmega328P is working as the memory storage of the Robot. Using its legs which are made of servo motors, the Robot can likewise move in a few stages.

Regions engaged in achieving this target which is elaborated here examine the servo motors, analyzing its turning effects, supportive system arrangement adjustment, generate dancing and walking algorithm, picking the extent of the ultrasonic sensor and impact on the servo horns close by the LED. This humanoid Robot (**Figure 1**) is about 17 cm in length and is very lightweight.

The coming segments are organized as follows: segment 2 depicts related works; segment 3 delineates Application. Segment 4 inspires working mechanism,

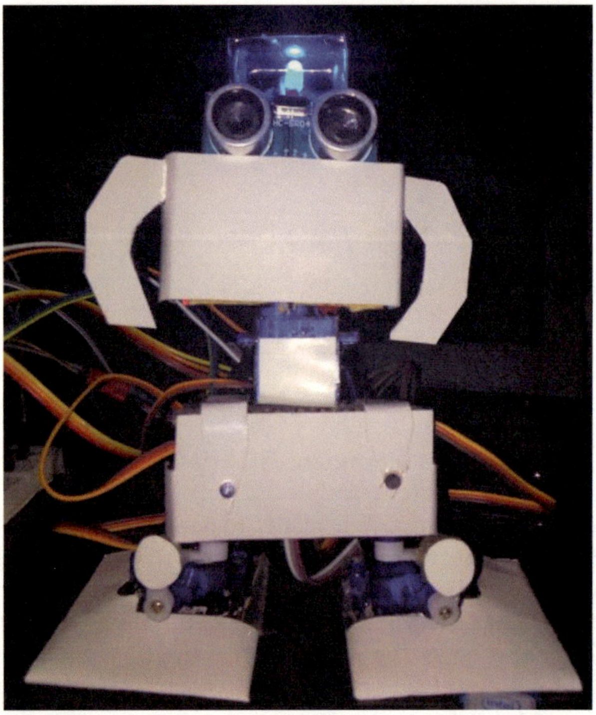

Figure 1.
Tiny humanoid strolling and moving robot.

area 5 shows programming mechanism and segment 6 illustrates Human-Robot Collaboration, Section 7 portrays hardware and financial requirement analysis and user opinion. Section 8 concludes this paper and references have been added towards the end of this paper.

2. Related works

Biped robots are the essential bit of humanoid robots, walking development examination is prioritized. Numerous explores happening on the creation of humanoid biped robot which is indistinguishable from an individual. Likewise, the robot relationship with individuals and headway of collaborative task are crucial perspectives that the humanoids are confronting. In these particular conditions, in the latest years, robot interactivity transformed into an essential area of examination.

Prakash Chandran, Mohit Jaswal, Dr. T.V.U Kiran Kumar and Mrs. Raji Pandurangan have introduced a biped humanoid (**Figure 2**) utilizing a microcontroller in their work distributed on International Journal of Science, Engineering and Technology Research (IJSETR). Here, the biped robot can detect sound and range distance. In this research work, they have referenced two different ways of designing humanoid robots. First is the regular spot technique where the Robot strolls in the course of action where the focal point of mass of Robot keeps moving to settle the position of the robot body. Another technique for implementing the change is using an accelerometer and gyro sensors. They chose the first strategy to adjust the structure. A sound sensor makes the Robot accept directions as voice orders. Such compound activities show utilization of open-source tasks, for example, Arduino [4].

Self-guided humanoid robot-Adhvik presented by Aditya Mishra, Ashutosh Shrivastav, Neha Maurya, S. Vamshi Krishna can perceive any red-shaded article before

Figure 2.
Humanoid robot (autonomous) using Arduino.

it alongside strolling and different functionalities. The Robot is fundamentally utilized for object location and wayfinding. This functionality has been refined using a remote camera. The camera divides the red object into multiple parts. The humanoid can explore three directions: Left, Right and Straight as indicated by the position of the ball. At the point when the camera catches the picture of the ball, it partitions the casing into three and afterward humanoid moves appropriately. MATLAB programming has been done to handle the picture. It includes the contribution of the remote module [5].

Numerous scientists are working and have designed visually impaired assistive gadgets and robots to give autonomous route facilities to the visually impaired. For instance, a smart way direction robot to help blinds portrayed in the paper of M F Razali, S F Toha and Z ZAbidin distributed on IEEE International Symposium on Robotics and Intelligent Sensors (IRIS 2015) utilizes a Fuzzy Logic Controller. A sensible, quick, and trustworthy arrangement is offered by the fuzzy logic approach, which is an advantage over the ordinary controller. The straightforward "IF-THEN" rule is expected to impact the activity in Fuzzy Logic [6].

Another tiny human structured robot that is distantly controlled by means of Bluetooth is represented in the research work of Ahmet Aksoz, Salim Engin, and Mahir Dursun in the Journal of Automation and Control Engineering Vol. 4, No. 3, June 2016. Their Robot utilizes the remote correspondence innovation that empowers it to be constrained by an Android application. In a far-off area, the Robot can move freely [7].

There are numerous kinds of exploration equipped for obliging and actualizing complex theories and algorithms for dynamic strolling, human interaction, navigation, human conduct simulation, artificial intelligence, visual recognition utilizing Image recognition [8]. Biped robots using these algorithms and techniques

Features	Other mentioned robots	Tiny Humanoid
Uses only four servos for strolling	3 robots among 5 have used four servos. Other 2 have used 6 or more.	Yes.
Detects obstacle and stops strolling	2 robots among 5 does this. Other 3 not defined.	Yes.
Gives notification using buzzer	None of the mentioned have used.	Yes.
Detects obstacle in 180 degree view	Detects only in front view.	Yes.
Can dance in several steps to entertain	None of the mentioned have done this.	Yes.
Gives notification using LED	None of the mentioned have done this.	Yes.

Table 1.
Comparison table of tiny humanoid with other robots.

are further developed and effective henceforth more costly and confounded. Our project likewise expects to actualize such controlling frameworks and complicated calculations in our humanoid in the forthcoming future. **Table 1** shows the comparison between Tiny Humanoid with other robots.

3. Application

Advancement of impelled humanoids is used in implementation in a tremendous extent of areas, for instance, the development industry, home applications, incitement, social protection, sport, space, education. Our small walking biped robot has features; for example, it can recognize article close to it and quit strolling movement illuminating the LED as an identifier at its top. The following situations describe the incredible use of the features.

The Robot can add to bioengineering by helping the outwardly disabled patient. Outwardly impeded individual faces troubles in exploring starting with one spot then onto the next evading object. The Robot can function as a visually impaired assistive robot all things considered. This Robot can identify objects before it utilizes the ultrasonic sensor and stops strolling by initializing the servo motors' turning estimations to its primary value alarming the visually impaired individual that there is an article close by.

The Robot additionally has features that can accelerate commercialization. People these days show more fascination towards robots. As the Robot is able to dance in a few steps, it very well may be used before amusement parks and restaurants to invite and welcome individuals with fascinating movements. Thus they can have more customers attracted to their place and accelerate business.

Children and kids are turning into the Smartphone focused step by step. It is damaging to their eyesight as they are consistently playing with a Smartphone. For them, the Robot can be an incredible apparatus to play with as it can dance and entertain them, distracting them from Smartphone.

The Robot's object recognition capacity can be utilized to guarantee a specific spot or item's safety ensures that not even a solitary item draws close to it.

Nonetheless, there are multiple purposes and utilization of this sort of humanoid Robot. The humanoid can be upgraded with different highlights, for example, talking, ignoring objects, tuning in to voice order that will without a doubt improve its Application.

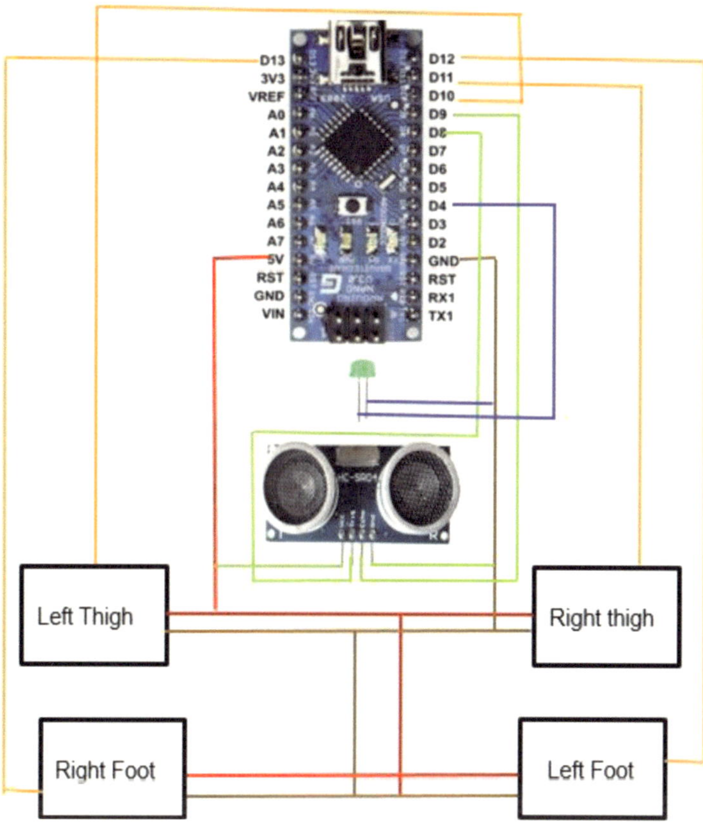

Figure 3.
Connection structure of tiny humanoid.

4. Working mechanism

From **Figure 3** we can without much of a stretch comprehend the Robot's association anatomy and how the ultrasonic sensor, servo motors [1] and the LED are associated with the Arduino board (Atmega 328P).

Each leg has been utilized using two servos. Servos are joined in a position, that the thigh servo's movement will influence the direction of the foot servo; however, the thigh servo motors are attached to the structure and are not influenced by the foot servo motor's movement. As the thigh servos are connected to the body, when the foot servo moves, it makes the route of the entire robot body alongside the thigh servo. The LED functions by coworking with the sensor through code.

As the appropriate strolling of the Robot relies upon the turning estimations of the servo motor horns, it is must be cautious while connecting the servos realizing the underlying pivoting values. In this way, a legitimate comprehension of servo motor rotation impact on its strolling is likewise significant. Paths need to be connected in a cautious manner, such as in servos, wiring positives to voltage and negatives to the ground; otherwise, it may damage the equipment.

Biped's strolling and objects recognizing steps:

- The servo of the right foot ascents up and the servo motor right thigh turns to make the servos of the foot to approach.

- Similarly, the servo of the leftmost thigh will pivot to the contrary side a piece to help the rightmost servo motors go ahead.
- After the right servo motors is put ahead, the leftmost foot servo motor raises and the leftmost thigh servo turns more prominent than the right foot servo and pushes a stride forward.
- Same time the right thigh servo motor does likewise as the leftmost thigh servo motor did.
- In that case, if an item is identified by the ultrasonic sensor, the drove lights and the humanoid set the servo motors to the initial state.
- This step is recalled while the microcontroller is controlled.

In this cycle, the Robot explores starting with one spot then onto the next.

5. Programming technique

Our methodology towards the coding began with investigating the impact of rotating values on servo motors. The left thigh and right thigh were at first set to 90 degrees. The left foot was set to 0 and the right foot was placed to 180 degrees as shown in **Figure 4**.

Programming steps:

1. Define four servo objects to operate the legs.

```
#include <Servo.h>
Servo rightfoot;
Servo rightthigh;
Servo leftfoot;
Servo leftthigh;
```

2. Initialize ports to the defined objects.

3. Mention the primary position of the servo motors.

```
rightfoot.attach(10);
rightthigh.attach(7);
leftfoot.attach(5);
leftthigh.attach(11);

leftfoot.write(0);
leftthigh.write(90);
rightthigh.write(90);
rightfoot.write(180);
```

4. Stroll by turning the servo horns in estimated angles.

```
rightfoot.write(170);
rightfoot.write(160);
```

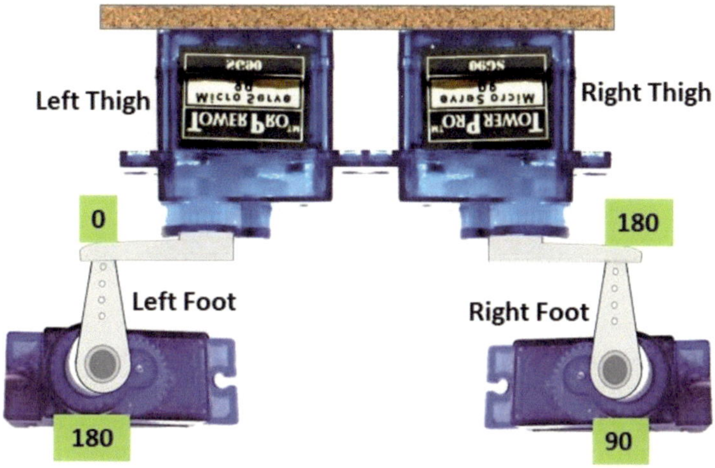

Figure 4.
Turning values of servo horns.

5. Identify if there is an obstacle found within a distance of less than 30.

```
if ((distance <= 30))
{
  digitalWrite(led, HIGH);
  leftfoot.write(0);
  leftthigh.write(90);
  rightthigh.write(90);
  rightfoot.write(180);
}
else if (distance > 30)
{
  digitalWrite(led, LOW);
}
```

6. If found, initialize servo horns to their initial position and illuminate LED.

```
digitalWrite(trigPin, HIGH);
delayMicroseconds(1000);
digitalWrite(trigPin, LOW);
```

7. If not found, repeat step 4, keeping the LED low.

Few lines are shown here as an example, this is not the complete code. One will need to generate the full code by understanding the algorithm appropriately. A suggestion is not to copy the above code but to understand it.

The above steps are programmed on the Arduino integrated development environment (IDE). It is a cross-stage application written in the Java programming language [9].

6. Human-robot collaboration

In human-robot cooperation, Both contribute their particular capacities. The robots play out the actual work. The human operator controls and screens. The Robot helps the visually impaired and hard of hearing human to move securely. This implies Robot does not supplant the human, however supplements his abilities and soothes him of difficult errands. They uphold a user as a partner on account of seeing obstacles close to them. Using the Bluetooth module, the user will be able to connect to the Robot and control its walking and view with the help of a mobile platform. The human operator will be able to move the Robot's head to the left and right using the platform.

7. Apparatus and monetary requirement analysis and user feedback

In this project, we did not add AA cells, yet it is important to add AA cells to control the microcontroller instead of fueling it with USB so the humanoid can stroll freely. **Figure 5** shows the wiring outline of utilizing the AA cells.

The Robot can be designed using Arduino UNO too. In that case, one can attach the Arduino UNO to the card body of the Robot. A smaller breadboard can also be attached at the back of the robot body. Attaching Arduino UNO and a small

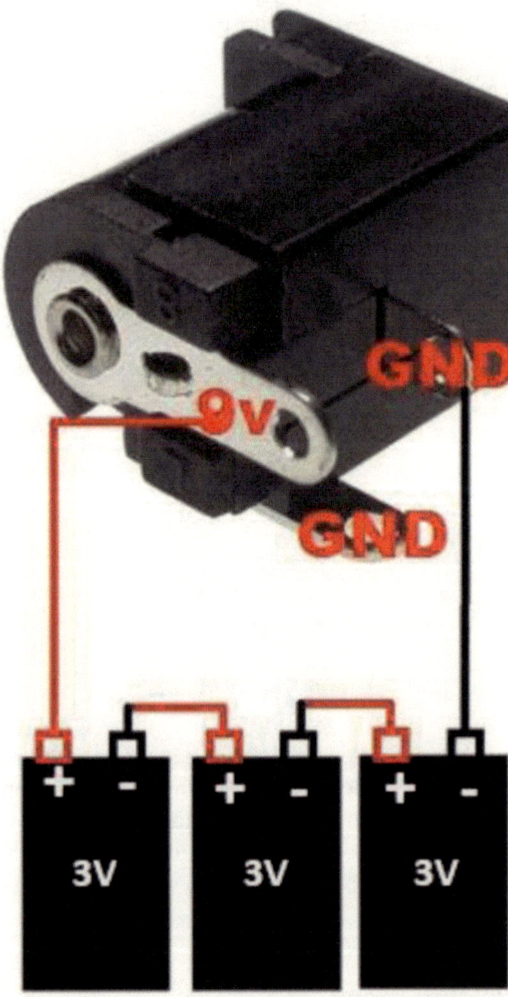

Figure 5.
Wiring diagram of AA cells.

Apparatus	Quantity	Price (BDT)
Arduino Nano V (3.0)	1	350
Micro servo (SG90)	4	135 * 4
Ultrasonic Sonar sensor (HC-SR04)	1	100
LED	1	2
Bread board	1	85
AA cells (Optional)	1	~60
Jumper wires	3 sets	20*3
Total: 1,197		

Table 2.
Apparatus with quantity and price.

breadboard to the robot body may restrict the Robot from walking properly. As it will increase the weight of the Robot, it is wise to use Arduino nano.

The Arduino Nano is an Arduino UNO contracted into a little profile, making it advantageous for restricted spaces and tasks that may have to reduce weight at every possible opportunities. Often, Arduino Nano is the least expensive choice accessible making the projects cost-effective. One should aim to decrease the use of long wires and reduce the weight of the Robot for better performance. Using a Bluetooth module will make it easier to achieve less weight.

Table 2 shows the significant apparatus and their price. Some other necessary apparatus incorporates glue gun, knife, and slices of cardboard to make the anatomy of the tiny humanoid.

As the Robot is grown mainly for outwardly weakened individuals, we went to Bangladesh Visually Impaired People's Society (BVIPS) [10] and Green Disabled Foundation (GDF) [11] to get user opinion. BVIPS is a deliberate association of visually impaired individuals in Bangladesh. The important opinion for this Robot is gathered from multiple association individuals mentioned in **Table 3**.

As the humanoid is developed mainly for apparently debilitated people, we have looked into Bangladesh Visually Impaired People's Society (BVIPS) [10] and Green

Institution	Participants
IER Dept, Dhaka University	22
BVIPS	35
GDF	19

Table 3.
Number of participants from different institution.

Benchmark	Compensation
Blind assistant	61
Amusement	43
Obstacle awareness	70

Table 4.
Blind people feedback.

Disabled Foundation (GDF) [11] to get client input. In Bangladesh, BVIPS is a purposeful relationship of outwardly disabled people.

Table 4 presents the feedback of the Robot that we have gotten from outwardly weakened individuals dependent on robot associate conduct, Entertainment and impediment location highlight.

8. Conclusion

In this chapter, we depicted the component to develop a simple humanoid robot to convert it to a gadget for future exploration and adjustment. This cost-effective tiny humanoid Robot can be used to assist visually impaired and hard of hearing at the same time. It can also be a source of entertainment and can provide security. Future researchers can contribute to the advancement of this Robot by adding features like voice recognition, control system from smartphones, decision-making capabilities using AI.

Regardless, the point is to manufacture a more extraordinary humanoid robot that will serve our overall population and make our life easier.

Author details

Amrita Ganguly and Bijan Paul*
University of Liberal Arts Bangladesh, Dhaka, Bangladesh

*Address all correspondence to: bijan.paul@ulab.edu.bd

References

[1] A. Bhargava and A. Kumar, "Arduino controlled robotic arm," 2017. International conference of Electronics, Communication and Aerospace Technology (ICECA), Coimbatore, India, 2017, pp. 376-380, doi: 10.1109/ICECA.2017.8212837.

[2] A. Dimitrov and D. Minchev, "Ultrasonic sensor explorer," 2016 19th International Symposium on Electrical Apparatus and Technologies (SIELA), Bourgas, Bulgaria, 2016, pp. 1-5, doi: 10.1109/SIELA.2016.7542987.

[3] L. Kamelia, M. R. Effendi and D. F. Pratama, "Integrated Smart House Security System Using Sensors and RFID," 2018 4th International Conference on Wireless and Telematics (ICWT), Nusa Dua, Bali, Indonesia, 2018, pp. 1-5, doi: 10.1109/ICWT.2018.8527803.

[4] P. Chandran, M. Jaswal, K. Kumar, R. Pandurangan. Autonomous Biped Robot Using Arduino. Internation Journal of Science, Engineering and Technology Research(IJSETR). volume 4, Issue 4, April 2015.

[5] Mishra, A., Shrivastav, A., Maurya, N., Vamshi Krishna, S. (2019). Self GuidedAdhvik Humanoid Robot. [internet] ijstr.org. Available from: http://www.ijstr.org/final-print/june2014/Self-Guided-Adhvik-Humanoid-Robot.pdf [Accessed: 6- Jan- 2019].

[6] Razali, M., Toha, S. and Abidin, Z. Intelligent Path Guidance Robot for Visually Impaired Assistance. In: X. Zeng editors. Procedia Computer Science. 2015. p. 330-335. DOI: https://doi.org/10.1016/j.procs.2015.12.303

[7] Aksoz, A., Engin, S., Dursun, M. (2019). The Implementation of Controlled Humanoid Robot with Android. [online] joace.org. Available at: http://www.joace.org/uploadfile/2015/1015/20151015022816477.pdf [Accessed 24- Jan- 2019].

[8] Hashemi, S., Moradinasab Mahdi Najmabadi, M. and Abdoshshah, S. (2019). [online] Ais.uni-bonn.de. Available from: http://www.ais.uni-bonn.de/humanoidsoccer/qualification/Iran_Fanavaran_TDP.pdf [Accessed 24 Jan. 2019].

[9] "Aduino IDE," Wikipedia, 07-Feb-2019. [Internet]. Available from: https://en.wikipedia.org/wiki/Arduino_IDE. [Accessed: 15-Feb-2019].

[10] mroberts (2017). Sightsavers homepage. [Internet] Sightsavers. Available from: https://www.sightsavers.org/.

[11] Racooncode.com. Green Disabled Foundation | RacoonCode. [online] Available from: http://racooncode.com/community-development/web-development/gdf-bangladesh [Accessed: 20- Dec- 2019].

Chapter 10

Optimization Based Dynamic Human Motion Prediction with Modular Exoskeleton Robots as Interactive Forces: The Case of Weight Lifting Motion

Hyun-Joon Chung

Abstract

The optimization-based dynamics model is formulated for the weight lifting motion with human and exoskeleton model as interactive force term in this chapter. In the optimization algorithm, the human motion is defined as variables so that the motion which we want to generate (box lifting motion in this case) can be predicted. The objective function or cost function is defined as performance measure which can be switched by developer. In this paper we use the summation of each joint torque square which is considered as the dynamic effort for the motion. Constraints are defined as joint limits, torque limits, hand position, dynamic balance, exoskeleton assistive points, etc. Interaction force form exoskeleton robot can be derived as generalized coordinates and generalized force which are related to inertial reference frame and human body frame. The results can show how effective the exoskeleton robots are according to their assistive force.

Keywords: Optimization algorithm, dynamic motion prediction, human modeling and simulation, exoskeleton robot, force interaction

1. Introduction

To design or to assess the exoskeleton robot, it is necessary to simulate human motion with interactive force from exoskeleton robot. Thus, we need the human modeling and simulation which method the interactive force from exoskeleton can be applied at. This is key motivation in this study and we formulate human motion simulation with modular exoskeleton model and predict human motion with interactive from modular waist and knee exoskeleton robot. To do that we formulate human motion with assistive force from exoskeleton robot. There are many good researches based on optimization techniques for simulation in different areas [1–4]. There are couple of human modeling and simulation software such as OpenSim [5]. By given motion OpenSim can generate muscle forces for that motion. Other is Santos which is optimization-based motion simulation so called predictive dynamics [6]. Some of human motion simulation is developed under optimization-based motion simulation [7, 8]. And, lifting motions are studied with optimization based technique as well [9, 10].

The modeling and simulation for human-exoskeleton is developed in [11]. In this study, we also use optimization-based motion prediction and simulation method with recursive Lagrange's equations of motion which is highly nonlinear. It provides predicted human motion as well as joint torques and ground reaction forces for weight lifting. This method also gives us pretty fast calculation time and accurate results.

2. Mechanical modeling

2.1 Modular exoskeleton robot

There are many researches about exoskeleton with human-robot cooperation [12]. Due to the wearability, convenience, comfort and easy portability, modular exoskeleton robot is becoming a trend in the industrial work environment nowadays such as construction site, heavy industry, medical care, logistics, maintenance, manufacturing process, etc. In this study, the modular means the modular type according to the body parts of human being. For example, shoulder modular exoskeleton, knee modular exoskeleton. Biggest merit of modular exoskeleton may be the bringing more comfortability rather than full-body exoskeleton robots. Of course, it is possible only in the industrial area. If we look for the purpose of rehabilitation in hospital, it may be different story. Also, once we narrow down the application area, modular exoskeleton robot can be lighter, have more simple structure and can be more compact. Some area, you do not need active exoskeleton robot, and just passive exoskeleton is fine. Also, modular exoskeleton robot is applicable either together or separate case by case. Furthermore, modular exoskeleton is more economical compare to the full-body exoskeleton robot. The following **Figure 1** shows the concept design of modular exoskeleton robot which we are currently developing. In this study, exoskeleton assistive force can be applied human body as an external force through the optimization process.

2.2 Human model

Human model is constructed using mechanical structure with revolute joints and rigid body. Thus, each body segment is connected with revolute joint. There are 49 degrees of freedoms (DOF) in each body joint. Such as knee is 1 DOF, shoulder is 3DOFs

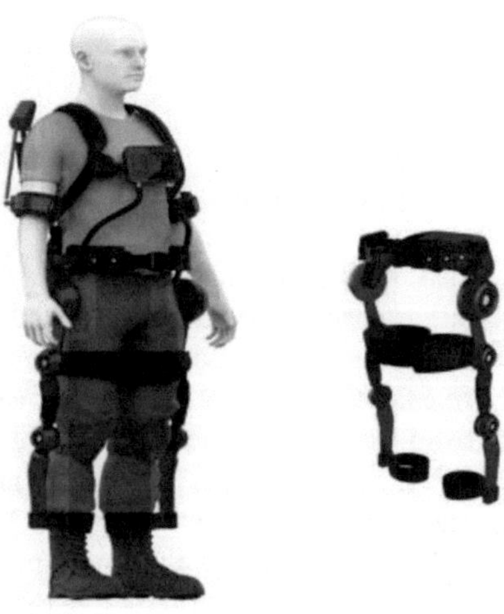

Figure 1.
The concept design of modular exoskeleton robot.

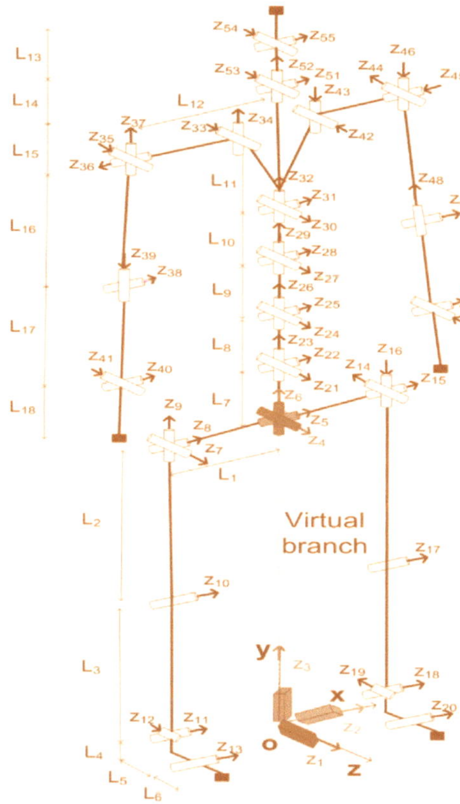

Figure 2.
Mechanical structure of human model.

etc. There are 6 DOFs for global translation and rotation for human system to the inertial reference frame. We put inertial reference frame at the point between the foot on the ground as human model is standing. We assume that body segment is rigid body and there is no muscle and tissue in this study. Therefore, it is assumed that all muscle force is converted to the joint torque. Also, we used GEBOD software to generate dynamic properties of body segment for example thigh, pelvis, and torso [13]. In this study we used 50 percentile male data which is representing average male size. **Figure 2** describes the mechanical structure of current human model to generate weight lifting motion with waist and knee modular exoskeleton robots. Each z-axis has DOF and transformation matrix is combined sequentially from inertial reference frame to the head, hands, and toes as a branch. The virtual branch depicts global DOFs which is mentioned in previous – 3 global translations and 3 global rotations. The torso part of the human model has 4 spine joints and there are total 12 DOFs. Then, it leads to the right arm branch, left arm branch, and head branch. Right arm and left arm branch has 4 joints and 9 DOFs respectively including clavicle joint. Head branch has 2 joints and 5 DOFs. Right leg and left leg has 4 joints and 7 DOFs respectively.

3. Kinematics and dynamics analysis

3.1 Kinematics

Denavit-Hartenberg (DH) method is used to analyze the kinematics of human motion [14]. In the Denavit-Hartenberg method any point ${}^{i}\mathbf{r}$ can be transferred to the global reference frame as ${}^{0}\mathbf{r}$ in **Figure 3** and it can be presented as Eq. (1).

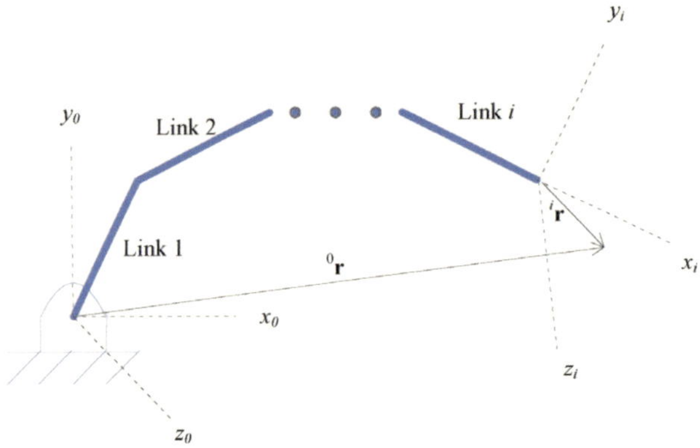

Figure 3.
Articulated chain.

$$ {}^0\mathbf{r} = {}^0\mathbf{T}_i {}^i\mathbf{r} \tag{1} $$

Transformation matrix ${}^0\mathbf{T}_i$ in Eq. (1) can be obtained as follows.

$$ {}^0\mathbf{T}_i = {}^0\mathbf{T}_1 {}^1\mathbf{T}_2 \cdots {}^{i-1}\mathbf{T}_i = \prod_{n=1}^{i} {}^{n-1}\mathbf{T}_n \tag{2} $$

$$ {}^{i-1}\mathbf{T}_i = \begin{bmatrix} \cos q_i & -\cos\alpha_i \sin q_i & \sin\alpha_i \sin q_i & a_i \cos q_i \\ \sin q_i & \cos\alpha_i \cos q_i & -\sin\alpha_i \cos q_i & a_i \sin q_i \\ 0 & \sin\alpha_i & \cos\alpha_i & d_i \\ 0 & 0 & 0 & 1 \end{bmatrix} \tag{3} $$

where the parameters q, α, a, d, are DH parameters. The q_i is the joint angle between the x_{i-1} axis and the xi axis about the z_{i-1} axis according to the right hand rule. The d_i is the distance between the origin of the i-1th coordinate frame and the intersection of the z_{i-1} axis with the x_i axis along the z_{i-1} axis. a_i is the distance between the intersection of the z_{i-1} axis with the x_i axis and the origin of the ith frame along the x_i axis. α_i is the angle between the z_{i-1} axis and the z_i axis about the x_i axis according to the right hand rule. In here, we use q_i as our generalized coordinates. As mentioned before, the inertial reference frame is located at the point between foot. The origin of inertial reference frame is O in the above **Figure 2**. Thus, all kinematic chain starts from origin of inertial reference frame. For the efficiency of calculation time, we use recursive way for kinematic information as follows:

$$ \mathrm{A}_i = \mathrm{A}_{i-1}\mathrm{T}_i \tag{4} $$

$$ \mathrm{B}_i = \dot{\mathrm{A}}_i = \mathrm{B}_{i-1}\mathrm{T}_i + \mathrm{A}_{i-1}\frac{\partial \mathrm{T}_i}{\partial q_i}\dot{q}_i \tag{5} $$

$$ \mathrm{C}_i = \dot{\mathrm{B}}_i = \mathrm{C}_{i-1}\mathrm{T}_i + 2\mathrm{B}_{i-1}\frac{\partial \mathrm{T}_i}{\partial q_i}\dot{q}_i + \mathrm{A}_{i-1}\frac{\partial^2 \mathrm{T}_i}{\partial q_i^2}\dot{q}_i^2 + \mathrm{A}_{i-1}\frac{\partial \mathrm{T}_i}{\partial q_i}\ddot{q}_i \tag{6} $$

where $\mathbf{A}_i$ is position matrix, $\mathbf{B}_i$ is velocity matrix, and $\mathbf{C}_i$ is acceleration matrix.

3.2 Dynamics

Once we obtain the kinematic information, we can use them to calculate dynamics of motion simulation. The equations of motion are derived from Lagrange's equation.

$$Q_i = \frac{d}{dt}\left(\frac{\partial L}{\partial \dot{q}_i}\right) - \frac{\partial L}{\partial q_i} \tag{7}$$

where L is Lagrangian which is $L = K - V$ (kinetic energy – potential energy), and q_i is the generalized coordinate. Total kinetic energy is derived as

$$K = \sum_{j=1}^{n} K_j = \frac{1}{2}\sum_{i=1}^{n} tr\left(\dot{\mathbf{A}}_i \mathbf{J}_i \dot{\mathbf{A}}_i^T\right). \tag{8}$$

where $\mathbf{A}_i$ is joint angle matrix and $\mathbf{J}_i$ is inertia matrix at i^{th} reference frame. Also, the potential energy for system can be given as

$$V = -\sum_{j=1}^{n} m_j \mathbf{g}^T \mathbf{A}_j {}^j\bar{\mathbf{r}}_j - \sum_{j=1}^{n} \mathbf{f}_k^T \mathbf{A}_j {}^k\mathbf{r}_f \delta_{jk} \tag{9}$$

where m_j is the mass of the body segment represented as j^{th} reference frame, $\mathbf{g}$ is gravity force vector, $\mathbf{f}_k$ is an external force which is defined in global reference frame and acting on the body segment expressed in k^{th} reference frame, ${}^k\mathbf{r}_f$ is the location of external force acting on the link expressed in the kth reference frame, δ_{jk} is Kronecker delta. Then, the equations of motion can be derived from above equations (Eqs. (7)–(9))

$$\tau_i = tr\left(\frac{\partial \mathbf{A}_i}{\partial q_i}\sum_{j=i}^{n} {}^i\mathbf{T}_j \mathbf{J}_j \ddot{\mathbf{A}}_j^T\right) - \mathbf{g}^T \frac{\partial \mathbf{A}_i}{\partial q_i}\sum_{j=i}^{n} m_j {}^i\mathbf{T}_j {}^j\bar{\mathbf{r}}_j - \mathbf{f}_k^T \frac{\partial \mathbf{A}_i}{\partial q_i}\sum_{j=i}^{n} {}^i\mathbf{T}_j {}^k r_f \delta_{jk} - \mathbf{G}_i \mathbf{A}_{i-1}\mathbf{z}_0 \tag{10}$$

where τ_i is the joint torque acting on the joint represented with generalized coordinate q_i, $\mathbf{G}_i$ is external moment which is defined in the global reference frame and $\mathbf{z}_0$ is $[0\ 0\ 1\ 0]^T$. Ground reaction force can be calculated using global force transformation the mechanical system of human body. Obtained joint torques are used to evaluate joint torque constraints and objective functions in later section.

4. Optimization formulation

Optimization based motion simulation in this chapter is performed according to the following process:

Step1: Prepare input data.
Step2: Function approximations for joint variables.
Step3: Kinematics analysis.
Step4: Dynamics analysis.
Step5: Objective function evaluation.
Step6: Constraints evaluation.
Step7: Print out if converge. Otherwise go back to step 2.

Kinematics and dynamics are covered in previous section, so we will discuss the joint variables, objective function, and constraints evaluation in this section.

4.1 Variables

To generate weight lifting motion, our optimization variable is joint angle profiles. These joint angle profiles are approximated using B-spline function approximation [15]. The control points in B-spline function approximation are updated in each iteration through the optimization process. Also, we used clamped B-spline which is that the starting point and end point of spline curve are matching to the control points. Following equation describes the B-spline approximation for the joint angle profiles q_i.

$$q_i(\mathbf{t}, \mathbf{p}) = \sum_{j=0}^{m} N_j(\mathbf{t}) p_j \tag{11}$$

where $\mathbf{t} = \{t_1, \ldots, t_s\}$ is knot vector, $\mathbf{p} = \{p_1, \ldots, p_m\}$ is the control points, N_j is the basis function of B-spline function approximation. Here, control points **P** becomes optimization variables.

4.2 Performance measure and objective function

We define the energy consuming (dynamic effort) for the given motion as the performance measure for weight lifting motion prediction and simulation. In this study, we use the joint torque square which is proportional to the mechanical energy. This mechanical energy is a reasonable criterion to minimize [4]. Then, it is formulated as objective function in optimization formulation as follows:

$$f = \int_0^T \left[\sum w_1 \tau_i^2 \right] dt \tag{12}$$

where w_i is the weighting parameters of each joint, τ_i is the joint torque of each joint. The joint torque is calculated from above dynamics equation (Eq. (10)).

4.3 Constraints

Constraints are the one of the motion control way in this optimization formulation and we used minimal set of constraints to generate motion in this formulation. The list of constraints are as follows:

1. Joint angle limits

2. Joint torque limits

3. Foot position and hand position

4. Stability condition

The joint angle limits and torque limits for the human motion are determined based on the literature [16–19]. Zero Moment Point(ZMP) method is used for the stability condition constraint [20]. Each constraint is formulated as follows accordingly:

$$\mathbf{q}^L \leq \mathbf{q}(t) \leq \mathbf{q}^U \tag{13}$$

$$\boldsymbol{\tau}^L \leq \boldsymbol{\tau}(t) \leq \boldsymbol{\tau}^U \tag{14}$$

$$\mathbf{r}_i(t) = \tilde{\mathbf{r}}_i \tag{15}$$

$$z_{zmp} \in \mathrm{FSR}, \quad x_{zmp} \in \mathrm{FSR} \tag{16}$$

4.4 Analytical gradients

The analytical gradients of each constraints and objective function are provided to the optimization solver. These analytical gradients improve the accuracy as well as calculation time for convergence in optimization process. Analytical gradients for constraints and objective function are as follows:

$$\frac{\partial C}{\partial x_1} = \frac{\partial C}{\partial q_1}\frac{\partial q_1}{\partial x_1} + \frac{\partial C}{\partial q_2}\frac{\partial q_2}{\partial x_1} + \cdots + \frac{\partial C}{\partial q_n}\frac{\partial q_n}{\partial x_1} + \frac{\partial C}{\partial \dot{q}_1}\frac{\partial \dot{q}_1}{\partial x_1} + \cdots + \frac{\partial C}{\partial \dot{q}_n}\frac{\partial \dot{q}_n}{\partial x_1} + \cdots + \frac{\partial C}{\partial \ddot{q}_n}\frac{\partial \ddot{q}_n}{\partial x_1}$$
$$\frac{\partial C}{\partial x_2} = \frac{\partial C}{\partial q_1}\frac{\partial q_1}{\partial x_2} + \frac{\partial C}{\partial q_2}\frac{\partial q_2}{\partial x_2} + \cdots + \frac{\partial C}{\partial q_n}\frac{\partial q_n}{\partial x_2} + \frac{\partial C}{\partial \dot{q}_1}\frac{\partial \dot{q}_1}{\partial x_2} + \cdots + \frac{\partial C}{\partial \dot{q}_n}\frac{\partial \dot{q}_n}{\partial x_2} + \cdots + \frac{\partial C}{\partial \ddot{q}_n}\frac{\partial \ddot{q}_n}{\partial x_2}$$
$$\vdots$$
$$\frac{\partial C}{\partial x_m} = \frac{\partial C}{\partial q_1}\frac{\partial q_1}{\partial x_m} + \frac{\partial C}{\partial q_2}\frac{\partial q_2}{\partial x_m} + \cdots + \frac{\partial C}{\partial q_n}\frac{\partial q_n}{\partial x_m} + \frac{\partial C}{\partial \dot{q}_1}\frac{\partial \dot{q}_1}{\partial x_m} + \cdots + \frac{\partial C}{\partial \dot{q}_n}\frac{\partial \dot{q}_n}{\partial x_m} + \cdots + \frac{\partial C}{\partial \ddot{q}_n}\frac{\partial \ddot{q}_n}{\partial x_m} \tag{17}$$

$$\frac{\partial \tau}{\partial x_k} = \frac{\partial \tau}{\partial q_j}\frac{\partial q_j}{\partial x_k} + \frac{\partial \tau}{\partial \dot{q}_j}\frac{\partial \dot{q}_j}{\partial x_k} + \frac{\partial \tau}{\partial \ddot{q}_j}\frac{\partial \ddot{q}_j}{\partial x_k} \tag{18}$$

where C represents constraints, q_i is joint angle profile, x_i is control point of B-spline function approximation, τ is joint torque profile. The analytical gradients of each constraint are calculated in the form of Eq. (17). For the analytical gradient of objective function is calculated by using Eqs. (12) and (18).

5. Numerical experiments

5.1 Weight lifting motion

Figure 4 depicts weight lifting motion. The weight is defined as W and it is applied as and external load in dynamics equilibrium equations of motion. Weight object is virtual and only hand location is guided by position constraints. Foot is located on the ground and weight is moving up vertically in 0.15 m from 0.5 m above ground and away from heal position by 0.25 m as shown in **Figure 4**. Dynamic stability constraint which is ZMP are imposed on the foot support region area. Joint angle limits and joint torque limits are imposed based on the literature review. It is assumed that the assistive moment is applied hip joint and knee joint which axes are parallel to the horizontal axis of inertial reference frame.

5.2 Simulation setting

For the optimization, sequential quadratic programming (SQP) is adopted. The sequential quadratic programming is very effective for the large scale nonlinear constrained optimization problem. The commercial software SNOPT is used which is well known as the effectiveness for large scale nonlinear problem SQP solver [21]. Total number of variables for optimization is 330 and total number of constrains is 1,766. The weighting parameters in objective function are set to equally in current study. Numerical experiment is performed while the weights are set to 15 kg and 30 kg. Also, different assistive moments from exoskeleton robot are tested in the experiment. The postprocessing for animation and snapshot was performed using Commercial software MATLAB.

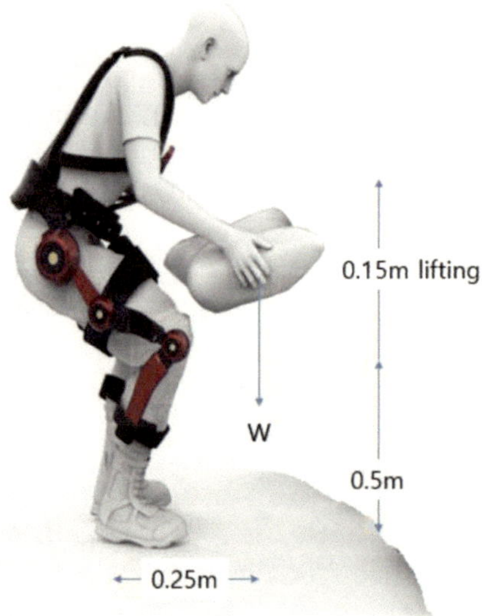

Figure 4.
Weight lifting motion with waist and knee exoskeleton robots.

6. Results

The simulation results are shown in following **Figure 5**. The assistive force from exoskeleton robot is tested from 10 N/m to 50 N/m for waist and knee exoskeleton. Then, we checked energy consumption for the lifting motion from the simulation. The results are **Table 1** and **Figure 6**. As shown in the **Table 1**, it is obvious that the exoskeleton robot reduces the total energy consumption of weight lifting motion. Most energy minimized case for each 15 kg and 30 kg lifting case is written in italic in the table. In some case, the total energy is more than no assistive force case.

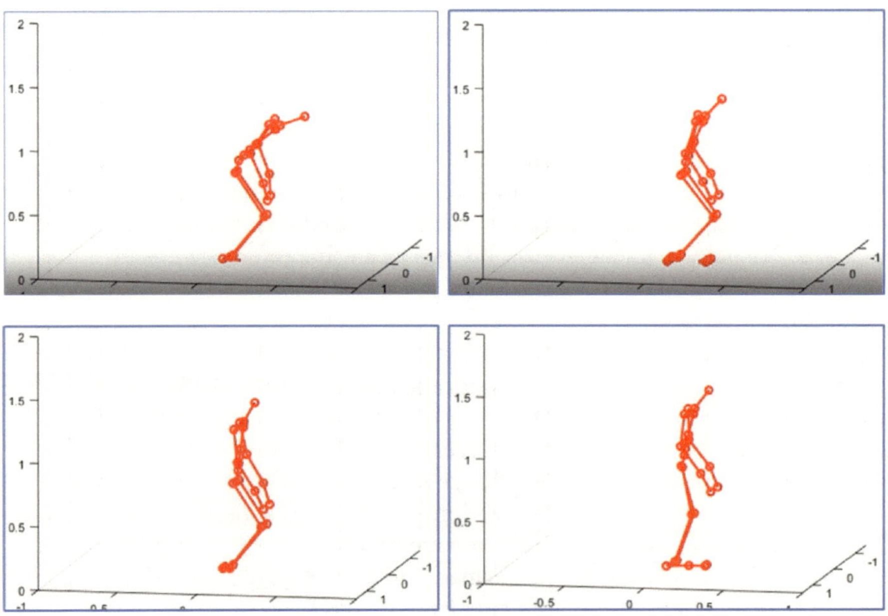

Figure 5.
Simulation snapshot for weight lifting motion.

Lifting weight	Exo assistive force (N/m)		Objective function value (energy consumption)	Lifting weight	Exo assistive force (N/m)		Objective function value (energy consumption)
	waist	knee			waist	knee	
15 kg	0	0	3.3260043054–002	30 kg	0	0	6.6957529363–002
	10	10	3.0190845608–002		10	10	4.6085529919–002
	20	10	2.7882680491–002		20	10	5.8372796346–002
	30	10	2.6598938935–002		30	10	5.3928838227 002
	40	10	2.6341574623–002		40	10	5.0509039531–002
	50	10	2.7108828860–002		50	10	4.8111291575–002
	10	20	2.4055151982–002		10	20	4.5952969619–002
	20	20	2.7788637489–002		20	20	4.5957540847–002
	30	20	2.6149510912–002		30	20	5.4008290188–002
	40	20	2.5534968466–002		40	20	5.0232961805–002
	50	20	2.5946755651–002		50	20	4.7482075385–002
	10	30	2.3808412818–002		10	30	4.6126512561–002
	20	30	2.3808412818–002		20	30	4.3911156137–002
	30	30	2.6053418696–002		30	30	4.4983526018–002
	40	30	2.5083330659–002		40	30	5.0310773358–002
	50	30	2.5137987323–002		50	30	4.7204074709–002
	10	40	2.4208434896–002		10	40	12.775145795–002
	20	40	2.2956322073–002		20	40	4.6699616710–002
	30	40	*2.2447356818–002*		30	40	4.6284398370–002
	40	40	2.4985189230–002		40	40	*4.2840986427–002*
	50	40	2.4684139732–002		50	40	4.7280245854–002
	10	50	2.4132838565–002		10	50	16.305738815–002
	20	50	2.3243927837–002		20	50	10.469578598–002
	30	50	*2.2435740271–002*		30	50	4.7176051325–002
	40	50	*2.1731970674–002*		40	50	*4.3894974761–002*
	50	50	2.4583949090–002		50	50	*4.2854122138–002*

Table 1.
Numerical experiment results for weight lifting motion.

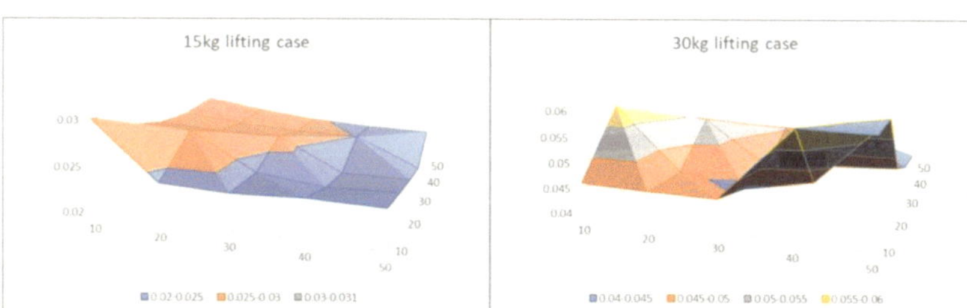

Figure 6.
The contour plot of numerical experiment results.

It might be the assistive force is bothering the balance mechanism of human body while weight lifting motion so human uses more energy to recover the balance back.

7. Conclusions

We have studied optimization-based motion simulation with modular waist and knee exoskeleton robot as an assistive force. We used Denavit-Hartenberg method for kinematics, Lagrange's equations of motion with external force and moment term, B-spline function approximation. In motion simulation, the performance measure is mechanical energy which is presented as the summation of joint torque squares. Minimal constraints are applied such as joint angle limits, torque limits, dynamic balance, and hand/foot positions., the optimization process find out the minimized energy consumed motion under the assistive forces from the modular waist and knee exoskeleton robots which are applied during the weight lifting motion.

This method provides unique feature with human-exoskeleton modeling and simulation area. It can give us predictive motion of human so that the exoskeleton parameters are adjusted based on the predicted motion simulation. Also, human motion can be generated automatically for the control algorithm of robot to collaborate with human. It can be used as evaluation and assessment tool for the design parameters of exoskeleton robot development in any given tasks according to human factors. Furthermore, this can reduce the development cost of exoskeleton because not many prototypes are necessary and provides safe design and test process during the exoskeleton development procedure. Of course, the musculoskeletal model should be developed for more accurate calculation of human factors and it will be remained as future works.

Acknowledgements

This research was financially supported by the Institute of Civil Military Technology Cooperation funded by the Defense Acquisition Program Administration and Ministry of Trade, Industry and Energy of Korean government under grant No. 19-CM-GU-01.

Author details

Hyun-Joon Chung
Korea Institute of Robotics and Technology Convergence, Pohang, The Republic of Korea

*Address all correspondence to: hjchung@kiro.re.kr

References

[1] A. Eriksson, "Optimization in Target Movement Simulations," Comput. Meth. in Appl. Mech. Eng., Vol. 197, pp. 4207–4215, 2008.

[2] G. Bessonnet, J. Marot, P. Seguin, and P. Sardain, "Parametric-Based Dynamic Synthesis of 3D-Gait," Robotica, Vol. 28, No. 4, pp. 563–581, 2010.

[3] C. L. Bottasso, B. I. Prilutsky, A. Croce, E. Imberti, and S. Sartirana, "A Numerical Procedure for Inferring From Experimental Data the Optimization Cost Functions Using a Multibody Model of the Neuro-Musculoskeletal System," Multibody Sys. Dyn., Vol. 16, No. 2, pp. 123–154, 2006.

[4] L. Roussel, C. Canuda-de-Wit, and A. Goswami, "Generation of Energy Optimal Complete Gait Cycles for Biped Robots," Proceedings of the IEEE International Conference on Robotics and Automation, Leuven, Belgium, Brussels, May 16-20, Vol. 3, pp. 2036-2041, 1998.

[5] S. L. Delp, F. C. Anderson, A. S. Arnold, P. Loan, A. Habib, C. T. John, E. Guendelman, and D. G. Thelen, OpenSim: Open-Source Software to Create and Analyze Dynamic Simulations of Movement, IEEE Transactions on Biomedical Engineering, Vol. 54, no. 11, 2007.

[6] Y. Xiang, H. J. Chung, J. H. Kim, R. Bhatt, S. Rahmatalla, J. Yang, T. Marler, J. S. Arora, and K. Abdel-Malek, "Predictive Dynamics: An Optimization-Based Novel Approach for Human Motion Simulation," Struct. Multidiscip. Optim., Vol. 41, No. 3, pp. 465–479, 2010.

[7] H. J. Chung, J. S. Arora, K. Abdel-Malek, and Y. Xiang, "Dynamic optimization of human running with analytical gradients," Journal of Computational and Nonlinear Dynamics, Vol. 10, no. 2, pp. 021006-1 – 021006-18, 2015.

[8] H. J. Chung, Y. Xiang, J. S. Arora, and K. Abdel-Malek, "Optimization-based dynamic 3D human running prediction: effects of feet location and orientation," Robotica, Vol. 33, no. 2, pp. 413-435, 2015.

[9] J. Song, X. Qu, C. Chen, "Lifting Motion Simulation Using a Hybrid Approach", Ergonomics, Vol. 58, No. 9, pp. 1557-1570, 2015.

[10] R. Zaman, Y. Xiang, J. Cruz, J. Yang, "Two-Dimensional Versus Three-dimensional Symmetric Lifting Motion Prediction Models: A Case Study". Journal of Computing and Information Science in Engineering. Vol. 21, No. 4, pp. 044501 (7 pages), 2021.

[11] M. Khamar, M. Edrisi, M. Zahiri, "Human-exoskeleton Control Simulation, Kinetic and Kinematic Modeling and Parameters Extraction," MethodsX, Vol. 6, pp. 1838/1846, 2019.

[12] H. D. Lee, B. K. Lee, W. S. Kim, J. S. Han, K. S. Shin, C.S. Han, "Human-robot cooperation control based on a dynamic model of an upper limb exoskeleton for human power amplification," Mechatronics, Vol. 24, pp. 168-176, 2014.

[13] H. Cheng, L. Obergefell, A. Rizer, Generator of body (GEBOD) manual, AL/DF-TR-1994-0051, Armstrong Laboratory, Wright-Patterson Air Force Base, Ohio, 1994.

[14] J. Denavit and R. S. Hartenberg, "A Kinematic Notation for Lower-pair Mechanisms Based on Matrices," ASME Journal of Applied Mechanics, Vol. 22, pp. 215-221, 1955.

[15] K. E. Atkinson, Introduction to Numerical Analysis 2nd edition. John Wiley & Sons, Inc., 1988.

[16] C. C. Norkin, and D. J. White, Measurement of Joint Motion: A Guide to Goniometry, 3rd Edition. Philadelphia: F. A. Davis Co, 2003.

[17] T. D. Cahalan, M. E. Johnson, and E. Y. Chao, "Quantitative Measurements of Hip Strength in Different Age Groups," Clin. Orthop. Relat. Res., Vol. 246, pp. 136-145, 1989.

[18] T. W. Kaminski, D. H. Perrin, and B. M. Gansneder, "Eversion strength Analysis of Uninjured and Functionally Unstable Ankles," J. Athl. Train., Vol. 34, No. 3, pp. 239-245, 1999.

[19] S. Kumar, Y. Narayan, and D. Garand, "An Electromyographic Study of Isokinetic Axial Rotation in Young Adults," Spine J. Vol. 3, No. 1, pp. 46-54, 2003.

[20] M. Vukobratović, B. and Borovac, "Zero-moment point – thirty five years of its life," International Journal of Humanoid Robotics, Vol. 1, No. 1, 157-173, 2004.

[21] P. E. Gill, W. Murray, and M. A. Saunders, "SNOPT: An SQP Algorithm for Large-Scale Constrained Optimization," SIAM J. Optim., Vol. 12, pp. 979-1006, 2002.